M000284507

The contents of this publication are solely those of the authors and contributors, and not of the publisher, editor(s), or employees of Silver Rock Publishing. Silver Rock Publishing, its parent company, and employees disclaim all responsibility for any injury or adverse effects of any kind to persons or property resulting from any ideas, methods, instructions, or products referred to in this publication. We present this book solely as an historical document and do not condone or endorse violence of any kind. Quite to the contrary, Silver Rock Publishing urges anyone considering acts of violence as a solution to seek professional assistance immediately.

U.S. ARMY IMPROVISED MUNITIONS HANDBOOK

Department of the Army

SILVER ROCK PUBLISHING

Published by Silver Rock Publishing

ISBN (Paperback): 978-1-62654-267-9
ISBN (Hardback): 978-1-62654-268-6
ISBN (Spiral): 978-1-62654-269-3

Printed and bound in the United States of America

FOR OFFICIAL USE ONLY

IMPROVISED MUNITIONS

HANDBOOK

TABLE OF CONTENTS

FRANKFORD ARSENAL

Philadelphia   Pennsylvania

FOR OFFICIAL USE ONLY

For further information or additional inserts, contact:

    Commanding Officer
    Frankford Arsenal
    ATTN: SMUF A-U3100, Special Products Division
            Small Caliber Engineering Directorate
    Philadelphia, Pa. 19137

    Additional inserts will be made available as evaluation
tests are completed. Please notify the above agency of any
change of address so that you may receive them.

FOR OFFICIAL USE ONLY

INTRODUCTION

1. Purpose and Scope

In Unconventional Warfare operations it may be impossible or unwise to use conventional military munitions as tools in the conduct of certain missions. It may be necessary instead to fabricate the required munitions from locally available or unassuming materials. The purpose of this Manual is to increase the potential of Special Forces and guerrilla troops by describing in detail the manufacture of munitions from seemingly innocuous locally available materials.

Manufactured, precision devices almost always will be more effective, more reliable, and easier to use than improvised ones, but shelf items will just not be available for certain operations for security or logistical reasons. Therefore the operator will have to rely on materials he can buy in a drug or paint store, find in a junk pile, or scrounge from military stocks. Also, many of the ingredients and materials used in fabricating homemade items are so commonplace or innocuous they can be carried without arousing suspicion. The completed item itself often is more easily concealed or camouflaged. In addition, the field expedient item can be tailored for the intended target, thereby providing an advantage over the standard item in flexibility and versatility.

The Manual contains simple explanations and illustrations to permit construction of the items by personnel not normally familiar with making and handling munitions. These items were conceived in-house or, obtained from other publications or personnel engaged in munitions or special warfare work. This Manual includes methods for fabricating explosives, detonators, propellants, shaped charges, small arms, mortars, incendiaries, delays, switches, and similar items from indigenous materials.

2. Safety and Reliability

Each item was evaluated both theoretically and experimentally to assure safety and reliability. A large number of items were discarded because of inherent hazards or unreliable performance. Safety warnings are prominently inserted in the procedures where they apply but it is emphasized that safety is a matter of attitude. It is a proven fact that men who are alert, who think out a situation, and who take correct precautions have fewer accidents than the careless and indifferent. It is important that work be planned and that instructions be followed to the letter; all work should be done in a neat and orderly manner. In the manufacture explosives, detonators, propellants and incendiaries, equipment must be kept clean and such energy concentrations as sparks,

5

friction, impact, hot objects, flame, chemical reactions, and excessive pressure should be avoided.

These items were found to be effective in most environments; however, samples should be made and tested remotely prior to actual use of assure proper performance. Chemical items should be used as soon as possible after preparation and kept free of moisture, dirt, and the above energy concentrations. Special care should be taken in any attempt at substitution or use of items for purposes other than that specified or intended.

3. User Comments

It is anticipated that this manual will be revised or changed from time to time. In this way it will be possible to update present material and add new items as they become available. Users are encouraged to submit recommended changes or comments to improve this manual. Comments should be keyed to the specific page, paragraph, and line of the text in which changes are recommended. Reasons should be provided for each comment to insure understanding and complete evaluation. Comments should be forwarded directly to Commandant, United States Army, Special Warfare School, Fort Bragg, North Carolina 28307 and Commanding Officer, United States Army, Frankford Arsenal, SMUFA-J8000, Philadelphia, Pennsylvania 19137.

PLASTIC EXPLOSIVE FILLER

A plastic explosive filler can be made from potassium chlorate and petroleum jelly. This explosive can be detonated with commercial #8 or any military blasting cap.

MATERIAL REQUIRED	HOW USED
Potassium chlorate	Medicine Manufacture of matches
Petroleum jelly (Vaseline)	Medicine Lubricant
Piece of round stick	
Wide bowl or other container for mixing ingredients.	

PROCEDURE

1. Spread potassium chlorate crystals thinly on a hard surface. Roll the round stick over crystals to crush into a very fine powder until it looks like face powder or wheat flour.

2. Place 9 parts powdered potassium chlorate and 1 part petroleum jelly in a wide bowl or similar container. Mix ingredients with hands (knead) until a uniform paste is obtained.

7

NOTE: Store explosive in a waterproof container until ready to use.

FOR OFFICIAL USE ONLY

POTASSIUM NITRATE

Potassium nitrate (saltpeter) can be extracted from many natural sources and can be used to make nitric acid, black powder and many pyrotechnics. The yield ranges from .1 to 10% by weight, depending on the fertility of the soil.

MATERIALS	SOURCE
Nitrate bearing earth or other material, about 3-1/2 gallons (13-1/2 liters)	Soil containing old decayed vegetable or animal matter
	Old cellars and/or farm dirt floors
	Earth from old burial grounds
	Decayed stone or mortar building foundations
Fine wood ashes, about 1/2 cup (1/8 liter)	Totally burned whitish wood ash powder
	Totally burned paper (black)

Bucket or similar container, about
5 gallons (19 liters) in volume
(Plastic, metal, or wood)
2 pieces of finely woven cloth, each
slightly larger than bottom of
bucket
Shallow pan or dish, at least as
large as bottom of bucket
Shallow heat resistant container
(ceramic, metal, etc.)
Water - 1-3/4 gallons (6-3/4 liters)
Awl, knife, screwdriver, or other
hole producing instrument
Alcohol about 1 gallon (4 liters)
(whiskey, rubbing alcohol, etc.)
Heat source (fire, electric heater, etc.)
Paper
Tape

NOTE: Only the ratios of the amounts of ingredients are important. Thus, for twice as much potassium nitrate, double quantities used.

PROCEDURE:

1. Punch holes in bottom of bucket. Spread one piece of cloth over holes inside of bucket.

Bottom of bucket

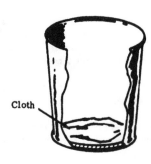

Cloth

2. Place wood ashes on cloth and spread to make a layer about the thickness of the cloth. Place second piece of cloth on top of ashes.

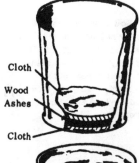

Cloth

Wood Ashes

Cloth

3. Place dirt in bucket.

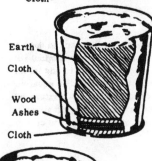

Earth

Cloth

Wood Ashes

Cloth

4. Place bucket over shallow container. Bucket may be supported on sticks if necessary.

Bucket

Stick

Shallow Container

9

5. Boil water and pour it over earth in bucket a little at a time.
Allow water to run through holes in bucket into shallow container. Be
sure water goes through all of the earth. Allow drained liquid to cool
and settle for 1 to 2 hours.

NOTE: Do not pour all of the water at once, since this may cause
stoppage.

6. Carefully drain off liquid into heat resistant container. Discard
any sludge remaining in bottom of the shallow container.

7. Boil mixture over hot
fire for at least 2 hours.
Small grains of salt will
begin to appear in the solution. Scoop these out as
they form, using any type
of improvised strainer
(paper, etc.).

8. When liquid has boiled down to
approximately half its original volume, remove from fire and let sit.
After half an hour add an equal volume of alcohol. When mixture is
poured through paper, small white
crystals will collect on top of it.

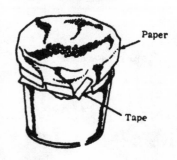

9. To purify the potassium nitrate, re-desolve the dry crystals in the
smallest possible amount of boiled water. Remove any salt crystals
that appear (Step 7); pour through an improvised filter made of several
pieces of paper and evaporate or gently heat the concentrated solution
to dryness.

10. Spread crystals on flat surface and allow to dry. The potassium
nitrate crystals are now ready for use.

IMPROVISED BLACK POWDER

Black powder can be prepared in a simple, safe manner. It may be used as blasting or gun powder.

MATERIAL REQUIRED:

Potassium nitrate, granulated, 3 cups (3/4 liter) (see Sect. I, No. 2)
Wood charcoal, powdered, 2 cups (1/2 liter)
Sulfur, powdered, 1/2 cup (1/8 liter)
Alcohol, 5 pints (2-1/2 liters) (whiskey, rubbing alcohol, etc.)
Water, 3 cups (3/4 liter)
Heat source
2 Buckets - each 2 gallon (7-1/2 liters) capacity, at least one of which
    is heat resistant (metal, ceramic, etc.)
Flat window screening, at least 1 foot (30 cm) square
Large wooden stick
Cloth, at least 2 feet (60 cm) square

NOTE: The above amounts will yield two pounds (900 grams) of black powder. However, only the ratios of the amounts of ingredients are important. Thus, for twice as much black powder, double all quantities used.

PROCEDURE:

1. Place alcohol in one of the buckets.

2. Place potassium nitrate, charcoal, and sulfur in the heat resistant bucket. Add 1 cup water and mix thoroughly with wooden stick until all ingredients are dissolved.

3. Add remaining water (2 cups) to mixture. Place bucket on heat source and stir until small bubbles begin to form.

CAUTION: Do not boil mixture. Be sure all mixture stays wet. If any is dry, as on sides of pan, it may ignite.

4.   Remove bucket from heat and
pour mixture into alcohol while
stirring vigorously.

Alcohol

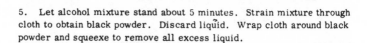

5.   Let alcohol mixture stand about 5 minutes. Strain mixture through
cloth to obtain black powder. Discard liquid. Wrap cloth around black
powder and squeexe to remove all excess liquid.

Cloth
Filter

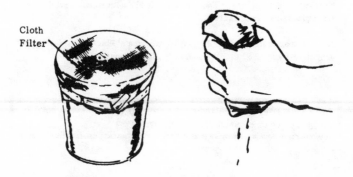

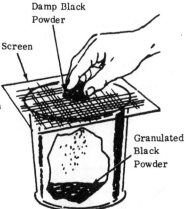

6. Place screening over dry bucket. Place workable amount of damp powder on screen and granulate by rubbing solid through screen.

NOTE: If granulated particles appear to stick together and change shape, recombine entire batch of powder and repeat steps 5 and 6.

7. Spread granulated black powder on flat dry surface so that layer about 1/2 inch (1-1/4 cm) is formed. Allow to dry. Use radiator, or direct sunlight. This should be dried as soon as possible, preferably in one hour. The longer the drying period, the less effective the black powder.

CAUTION: Remove from heat as soon as granules are dry. Black powder is now ready for use.

NITRIC ACID

Nitric acid is used in the preparation of many explosives, incendiary mixtures, and acid delay timers. It may be prepared by distilling a mixture of potassium nitrate and concentrated sulfuric acid.

MATERIAL REQUIRED:

Potassium nitrate (2 parts by volume)
Concentrated sulfuric acid (1 part by volume)
2 bottles or ceramic jugs (narrow necks are preferable)
Pot or frying pan
Heat source (wood, coal, or charcoal)
Tape (paper, electrical, masking, etc. but not cellophane)
Paper or rags

SOURCES:

Drug Store
Improvised (Section I, No. 2)
Motor vehicle batteries
Industrial plants

IMPORTANT: If sulfuric acid is obtained from a motor vehicle battery, concentrate it by boiling it until white fumes appear. DO NOT INHALE FUMES.

NOTE: The amount of nitric acid produced is the same as the amount of potassium nitrate. Thus, for 2 tablespoonsful of nitric acid, use 2 tablespoonsful of potassium nitrate and 1 tablespoonsful of concentrated sulfuric acid.

PROCEDURE:

1. Place dry potassium nitrate in bottle or jug. Add sulfuric acid. Do not fill bottle more than 1/4 full. Mix until paste is formed.

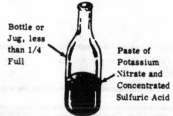

Bottle or Jug, less than 1/4 Full

Paste of Potassium Nitrate and Concentrated Sulfuric Acid

CAUTION: Sulfuric acid will burn skin and destroy clothing. If any is spilled, wash it away with a large quantity of water. Fumes are also dangerous and should not be inhaled.

2. Wrap paper or rags around necks of 2 bottles. Securely tape necks of bottles together. Be sure bottles are flush against each other and that there are no air spaces.

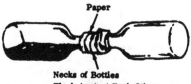

Paper

Necks of Bottles
Flush Against Each Other

3. Support bottles on rocks or cans so that empty bottle is <u>slightly</u> lower than bottle containing paste so that nitric acid that is formed in receiving bottle will not run into other bottle.

Tape Seal

Receiving
Bottle

Rocks or
Can Supports

4. Build fire in pot or frying pan.

5. Gently heat bottle containing mixture by moving fire in and out. As red fumes begin to appear periodically pour cool water over empty receiving bottle. Nitric acid will begin to form in the receiving bottle.

Water

15

CAUTION: Do not overheat or wet bottle containing mixture or it may shatter. As an added precaution, place bottle to be heated in heat resistant container filled with sand or gravel. Heat this outer container to produce nitric acid.

Heat Resistant Container
Filled with Sand or Gravel

6.   Continue the above process until no more red fumes are formed. If the nitric acid formed in the receiving bottle is not clear (cloudy) pour it into cleaned bottle and repeat Steps 2 - 6.

CAUTION: Nitric acid will burn skin and destroy clothing. If any is spilled, wash it away with a large quantity of water. Fumes are also dangerous and should not be inhaled.

Nitric acid should be kept away from all combustibles and should be kept in a sealed ceramic or glass container.

INITIATOR FOR DUST EXPLOSIONS

An initiator which will initiate common material to produce dust explosions can be rapidly and easily constructed. This type of charge is ideal for the destruction of enclosed areas such as rooms or buildings.

MATERIAL REQUIRED:

A flat can, 3 in. (8 cm) diameter and 1-1/2 in. (3-3/4 cm) high. A
    6-1/2 ounce Tuna can serves the purpose quite well.
Blasting cap
Explosive
Aluminum (may be wire, cut sheet, flattened can or powder
Large nail, 4 in. (10 cm) long
Wooden rod - 1/4 in. (6 mm) diameter
Flour, gasoline and powder or chipped aluminum

NOTE: Plastic explosives (Comp. C-4, etc.) produce better explosions than cast explosives (Comp. B, etc.).

PROCEDURE:

1. Using the nail, press a hole through the side of the Tuna can 3/8 to 1/2 inch (1 to 1-1/2 cm) from the bottom. Using a rotating and lever action, enlarge the hole until it will accommodate the blasting cap.

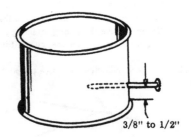

3/8" to 1/2"

2. Place the wooden rod in the hole and position the end of the rod at the center of the can.

3. Press explosive into the can, being sure to surround the rod, until it is 3/4 inch (2 cm) from top of the can. Carefully remove the wooden rod.

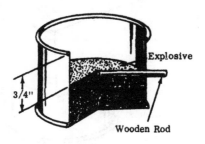

Explosive

3/4"

Wooden Rod

17

4.   Place the aluminum metal on top of the explosive.

5.   Just before use, insert the blasting cap into the cavity made by the rod. The initiator is now ready for use.

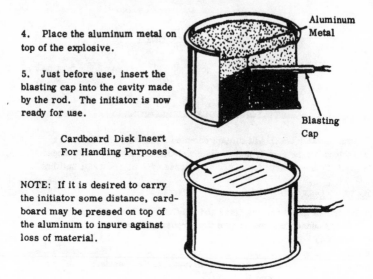

Aluminum Metal

Blasting Cap

Cardboard Disk Insert For Handling Purposes

NOTE: If it is desired to carry the initiator some distance, cardboard may be pressed on top of the aluminum to insure against loss of material.

HOW TO USE:

This particular unit works quite well to initiate charges of five pounds of flour, 1/2 gallon (1-2/3 liters) of gasoline or two pounds of flake painters aluminum. The solid materials may merely be contained in sacks or cardboard cartons. The gasoline may be placed in plastic-coated paper milk cartons, plastic or glass bottles. The charges are placed directly on top of the initiator and the blasting cap is actuated electrically or by fuse depending on the type of cap employed. This will destroy a 2,000 cubic feet enclosure (building 10 x 20 x 10 feet).

NOTE: For larger enclosures, use proportionately larger initiators and charges.

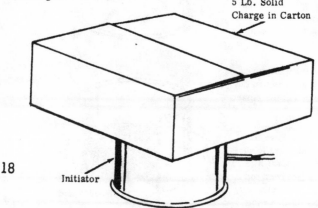

5 Lb. Solid Charge in Carton

Initiator

18

FERTILIZER EXPLOSIVE

An explosive munition can be made from fertilizer grade ammonium nitrate and either fuel oil or a mixture of equal parts of motor oil and gasoline. When properly prepared, this explosive munition can be detonated with a blasting cap.

MATERIAL REQUIRED:

Ammonium nitrate (not less than 32% nitrogen)
Fuel oil or gasoline and motor oil (1:1 ratio)
Two flat boards. (At least one of these should be comfortably held in the hand, i.e. 2 x 4 and 36 x 36.)
Bucket or other container for mixing ingredients
Iron or steel pipe or bottle, tin can or heavy-walled cardboard tube
Blasting cap
Wooden rod - 1/4 in. diameter
Spoon or similar measuring container

PROCEDURE:

1. Spread a handful of the ammonium nitrate on the large flat board and rub vigorously with the other board until the large particles are crushed into a very fine powder that looks like flour (approx. 10 min).

NOTE: Proceed with Step 2 as soon as possible since the powder may take moisture from the air and become spoiled.

2. Mix one measure (cup, tablespoon, etc.) of fuel oil with 16 measures of the finely ground ammonium nitrate in a dry bucket or other suitable container and stir with the wooden rod. If fuel oil is not available, use one half measure of gasoline and one half measure of motor oil. Store in a waterproof container until ready to use.

19

3. Spoon this mixture into an iron or steel pipe which has an end cap threaded on one end. If a pipe is not available, you may use a dry tin can, a glass jar or a heavy-walled cardboard tube.

NOTE: Take care not to tamp or shake the mixture in the pipe. If mixture becomes tightly packed, one cap will not be sufficient to initiate the explosive.

4. Insert blasting cap just beneath the surface of the explosive mix.

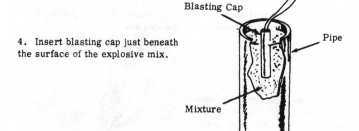

Blasting Cap

Pipe

Mixture

NOTE: Confining the open end of the container will add to the effectiveness of the explosive.

CARBON TET - EXPLOSIVE

A moist explosive mixture can be made from fine aluminum powder combined with carbon tetrachloride or tetrachloroethylene. This explosive can be detonated with a blasting cap.

MATERIAL REQUIRED:

Fine aluminum bronzing powder
Carbon tetrachloride
    or
tetrachloroethylene
Stirring rod (wood)
Mixing container (bowl, bucket, etc.)
Measuring container (cup, table-
    spoon, etc.)
Storage container (jar, can, etc.)
Blasting cap
Pipe, can or jar

SOURCE

Paint Store
Pharmacy, or fire extin-
    guisher fluid
Dry cleaners, Pharmacy

PROCEDURE:

1. Measure out two parts aluminum powder to one part carbon tetrachloride or tetrachloroethylene liquid into mixing container, adding liquid to powder while stirring with the wooden rod.

2. Stir until the mixture becomes the consistency of honey syrup.

CAUTION: Fumes from the liquid are dangerous and should not be inhaled.

21

3. Store explosive in a jar or similar water proof container until ready to use. The liquid in the mixture evaporates quickly when not confined.

NOTE: Mixture will detonate in this manner for a period of 72 hours.

HOW TO USE:

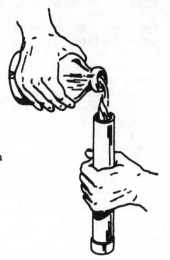

1. Pour this mixture into an iron or steel pipe which has an end cap threaded on one end. If a pipe is not available, you may use a dry tin can or a glass jar.

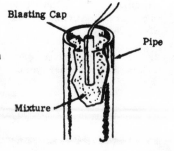

Blasting Cap

Pipe

Mixture

2. Insert blasting cap just beneath the surface of the explosive mix.

NOTE: Confining the open end of the container will add to the effectiveness of the explosive.

FERTILIZER AN-AL EXPLOSIVE

A dry explosive mixture can be made from ammonium nitrate fertilizer combined with fine aluminun powder. This explosive can be detonated with a blasting cap.

MATERIAL REQUIRED: SOURCE

Ammonium nitrate fertilizer   Farm or Feed Store
   (not less than 32% nitrogen)
Fine aluminum bronzing powder   Paint Store
Measuring container (cup, table-
   spoon, etc.)
Mixing container (wide bowl, can,
   etc.)
Two flat boards (one should be
   comfortably held in hand and
   one very large, i.e.
   2 x 4 and 36 x 36 in.)
Storage container (jar, can, etc.)
Blasting cap
Wooden rod - 1/4 inch diameter
Pipe, can or jar

PROCEDURE:

1. Method I - To obtain a low velocity explosive.

    a. Use measuring container to measure four parts fertilizer to one part aluminum powder and pour into the mixing container. (Example: 4 cups of fertilizer to 1 cup aluminum powder.)

    b. Mix ingredients well with the wooden rod.

2. Method II - To obtain a much higher velocity explosive.

    a. Spread a handful at a time of the fertilizer on the large flat board and rub vigorously with the other board until the large particles are crushed into a very fine powder that looks like flour (approx. 10 min per handful).

23

NOTE: Proceed with step b below as soon as possible since the powder may take moisture from the air and become spoiled.

    b.   Follow steps a and b of Method I.

3.   Store the explosive mixture in
a waterproof container, such as
glass jar, steel pipe, etc., until
ready to use.

HOW TO USE:

Follow steps 1 and 2 of "How To Use" in Section I, No. 7.

"RED OR WHITE POWDER" PROPELLANT

"Red or White Powder" Propellant may be prepared in a simple, safe manner. The formulation described below will result in approximately 2-1/2 pounds of powder. This is a small arms propellant and should only be used in weapons with 1/2 in. inside diameter or less, such as the Match Gun or the 7.62 Carbine, but not pistols.

MATERIAL REQUIRED:

Heat source (Kitchen stove or open fire)
2 gallon metal bucket
Measuring cup (8 ounces)
Wooden spoon or rubber spatula
Metal sheet or aluminum foil (at least 18 in. sq.)
Flat window screen (at least 1 ft. sq.)
Potassium nitrate (granulated) 2-1/3 cups
White sugar (granulated) 2 cups
Powdered ferric oxide (rust) 1/8 cup (if available)
Clear water, 3-1/2 cups

PROCEDURE:

1. Place the sugar, potassium nitrate, and water in the bucket. Heat with a low flame, stirring occasionally until the sugar and potassium nitrate dissolve.

2. If available, add the ferric oxide (rust) to the solution. Increase the flame under the mixture until it boils gently.

NOTE: The mixture will retain the rust coloration.

25

3. Stir and scrape the bucket sides occasionally until the mixture is reduced to one quarter of its original volume, then stir continuously.

4. As the water evaporates, the mixture will become thicker until it reaches the consistency of cooked breakfast cereal or homemade fudge. At this stage of thickness, remove the bucket from the heat source, and spread the mass on the metal sheet.

5. While the material cools, score it with the spoon or spatula in crisscrossed furrows about 1 inch apart.

6. Allow the material to air dry, preferably in the sun. As it dries, rescore it occasionally (about every 20 minutes) to aid drying.

7. When the material has dried to a point where it is moist and soft but not sticky to the touch, place a small spoonful on the screen. Rub the material back and forth against the screen mesh with spoon or other flat object until the material is granulated into small worm-like particles.

8. After granulation, return the material to the sun to dry completely.

NITRIC ACID/NITROBENZENE ("HELLHOFFITE") EXPLOSIVE

An explosive munition can be made from mononitrobenezene and nitric acid. It is a simple explosive to prepare. Just pour the mononitrobenzene into the acid and stir.

MATERIAL REQUIRED:	SOURCE:
Nitric acid	Field grade or 90% concentrated (specific gravity of 1.48)
Mononitrobenzene (also known as nitrobenzene)	Drug store (oil of mirbane) Chemical supply house Industries (used as solvent)
Acid resistant measuring containers	Glass, clay, etc.
Acid resistant mixing rod Blasting cap Wax Steel pipe, end cap and tape Bottle or jar	

NOTE: Prepare mixture just before use.

PROCEDURE:

1. Add 1 volume (cup, quart, etc.) mononitrobenzene to 2 volumes nitric acid in bottle or jar.

2. Mix ingredients well by stirring with acid resistant rod.

CAUTION: Nitric acid will burn skin and destroy clothing. If any is spilled, wash off immediately with large amount of water. Nitro-benzene is toxic; do not inhale fumes.

HOW TO USE:

1. Wax blasting cap, pipe and end cap.

2. Thread end cap onto pipe.

3. Pour mixture into pipe.

4. Insert and tape blasting cap just beneath surface of mixture.

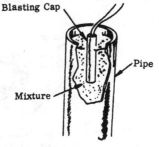

Blasting Cap

Pipe

Mixture

NOTE: Combining the open end of the pipe will add to the effectiveness of the explosive.

29

OPTIMIZED PROCESS FOR CELLULOSE/ACID EXPLOSIVES

An acid type explosive can be made from nitric acid and white paper or cotton cloth. This explosive can be detonated with a commercial #8 or any military blasting cap.

MATERIAL REQUIRED:	SOURCE:
Nitric Acid	Industrial metal processors, 90% concentrated (specific gravity of 1.48)
	Field grade (See Section I, No. 4)
White unprinted, unsized paper	Paper towels, napkins
Clean white cotton cloth	Clothing, sheets, etc.
Acid resistant container	Wax coated pipe or can, ceramic pipe, glass jar, etc.
	Heavy-walled glass containers
Aluminum foil or acid resistant material	Food stores
Protective gloves	
Blasting cap	
Wax	

PROCEDURE:

1. Put on gloves.

2. Spread out a layer of paper or cloth on aluminum foil and sprinkle with nitric acid until thoroughly soaked. If aluminum foil is unavailable, use an acid resistant material (glass, ceramic or wood).

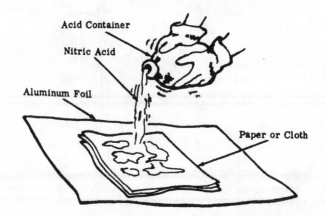

Acid Container

Nitric Acid

Aluminum Foil

Paper or Cloth

CAUTION: Acid will burn skin and destroy clothing. If any is spilled, wash it away with a large quantity of water. Do not inhale fumes.

3. Place another layer of paper or cloth on top of the acid-soaked sheet and repeat step 2 above. Repeat as often as necessary.

Rolled Sheets

4. Roll up the aluminum foil containing the acid-soaked sheets and insert the roll into the acid resistant container.

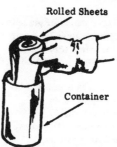

Container

NOTE: If glass, ceramic or wooden tray is used, pick up sheets with two wooden sticks and load into container.

5. Wax blasting cap.

Blasting Cap

6. Insert the blasting cap in the center of the rolled sheets. Allow 5 minutes before detonating the explosive.

31

METHYL NITRATE DYNAMITE

A moist explosive mixture can be made from sulfuric acid, nitric acid and methyl alcohol. This explosive can be detonated with a blasting cap.

MATERIAL REQUIRED:	SOURCES:
Sulfuric acid	Clear battery acid boiled until white fumes appear
Nitric acid	Field grade nitric acid (Section I, No. 4) or 90% conc. (sp. gr. of 1.48)
Methyl alcohol	Methanol Wood alcohol (not denatured alcohol) Anti-freeze (non-permanent)

Eyedropper or syringe with glass tube
Large diameter glass (2 qt.) jar
Narrow glass jars (1 qt.)
Absorbent (fine sawdust, shredded paper, shredded cloth)
Cup
Pan (3 to 5 gallon)
Teaspoon
Wooden stick
Steel pipe with end cap
Blasting cap
Water
Tray

PROCEDURE:

1. Add 24 teaspoons of sulfuric acid to 16-1/2 teaspoons of nitric acid in the 2 quart jar.

CAUTION: Acid will burn skin and destroy clothing. If any is spilled, wash it away with a large quantity of water. Do not inhale fumes.

2. Place the jar in the pan (3 to 5 gallon) filled with cold water or a stream and allow acid to cool.

3. Rapidly swirl the jar to create a whirlpool in the liquid (without splashing) while keeping the bottom portion of the jar in the water.

4. While continually swirling, add to mixture, 1/2 teaspoon at a time, 13-1/2 teaspoons of methyl alcohol, allowing mixture to cool at least one minute between additions.

CAUTION: If there is a sudden increase in the amount of fumes produced or if the solution suddenly turns much darker or begins to froth, dump solution in the water within 10 seconds. This will halt the reaction and prevent an accident.

5. After the final addition of methyl alcohol, swirl for another 30 to 45 seconds.

6. Carefully pour the solution into one of the narrow glass jars. Allow jar to stand in water for approximately 5 minutes until two layers separate.

7. With an eyedropper or syringe, remove top layer and carefully put into another narrow glass jar. This liquid is the explosive.

CAUTION: Explosive is shock sensitive.

8. Add an equal quantity of water to the explosive and swirl. Allow mixture to separate again as in step 6. The explosive is now the bottom layer.

9. Carefully remove the top layer with the eyedropper or syringe and discard.

10. Place one firmly packed cup of absorbent in the tray.

11. While stirring with the wooden stick, slowly add explosive until the mass is very damp but not wet enough to drip. Explosive is ready to use.

NOTE: If mixture becomes too wet, add more absorbent.

      If storage of explosive is required, store in a sealed container to prevent evaporation.

CAUTION: Do not handle liquid explosive or allow to contact skin. If this happens, flush away immediately with large quantity of water. Keep grit, sand or dirt out of mix.

34

HOW TO USE:

1. Spoon this mixture into an iron or steel pipe which has an end cap threaded on one end. If a pipe is not available, you may use a dry tin can or a glass jar.

2. Insert blasting cap just beneath the surface of the explosive mix.

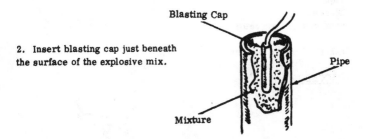

Blasting Cap

Pipe

Mixture

NOTE: Confining the open end of the container will add to the effectiveness of the explosive.

UREA NITRATE EXPLOSIVE

Urea nitrate can be used as an explosive munition. It is easy to prepare from nitric acid and urine. It can be detonated with a blasting cap.

MATERIAL REQUIRED:

Nitric acid, 90% conc. (1.48 sp. gr.)

Urine
2 one gallon heat and acid-resistant containers (glass, clay, etc.)
Filtering material

Aluminum powder (optional or if available)
Heat source
Measuring containers (cup and spoon)
Water
Tape
Blasting cap
Steel pipe and end cap(s)

SOURCE:

Field grade (Section I, No. 4) or industrial metal processors
Animals (including humans)

Paper towel or finely textured cotton cloth (shirt, sheet, etc.)
Paint stores

NOTE: Prepare mixture just before use.

PROCEDURE:

1. Boil a large quantity of urine (10 cups) to approximately 1/10 its volume (1 cup) in one of the containers over the heat source.

36

2. Filter the urine into the other container through the filtering material to remove impurities.

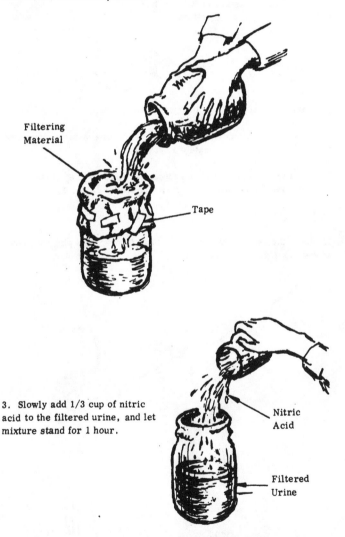

Filtering
Material

Tape

3. Slowly add 1/3 cup of nitric acid to the filtered urine, and let mixture stand for 1 hour.

Nitric
Acid

Filtered
Urine

CAUTION: Acid will burn skin and destroy clothing. If any is spilled wash it away with a large quantity of water. Do not inhale fumes.

37

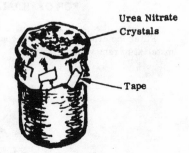

Urea Nitrate
Crystals

Tape

4. Filter mixture as in step 2.
Urea nitrate crystals will collect
on the paper.

5. Wash the urea nitrate by pouring water over it.

6. Remove urea nitrate crystals from the filtering material and allow
to dry thoroughly (approximately 16 hours).

NOTE: The drying time can be reduced to two hours if a hot (not
boiling) water bath is used. See Step 5 of Section I, No. 15.

HOW TO USE:

1. Spoon the urea nitrate crystals into an iron or steel pipe which has
an end cap threaded on one end.

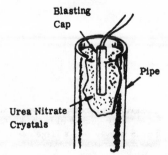

Blasting
Cap

Pipe

Urea Nitrate
Crystals

2. Insert blasting cap just beneath
the surface of the urea nitrate
crystals.

38

NOTES: This explosive can be made more effective by mixing with aluminum powder (can be obtained in paint stores) in the ratio of 4 to 1. For example, mix 1 cup of aluminum powder with 4 cups of urea nitrate.

Confining the open end of the container will add to the effectiveness of the explosive.

PREPARATION OF COPPER SULFATE (PENTAHYDRATE)

Copper sulfate is a required material for the preparation of TACC (Section I, No. 16).

MATERIAL REQUIRED:

Pieces of copper or copper wire
Dilute sulfuric acid (battery acid)
Potassium Nitrate (Section I, No. 2) or Nitric Acid, 90% conc. (1.48
    sp. gr.) (Section I, No. 4)
Alcohol
Water
Two 1 pint jars or glasses, heat resistant
Paper towels
Pan
Wooden rod or stick
Improvised Scale (Section VII, No. 8)
Cup
Container
Heat source
Teaspoon

PROCEDURE:

1. Place 10 grams of copper pieces into one of the pint jars. Add 1 cup (240 milliliters) of dilute sulfuric acid to the copper.

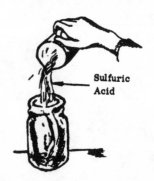

Sulfuric Acid

2. Add 12 grams of potassium nitrate or 1-1/2 teaspoons of nitric acid to the mixture.

Nitric Acid or Potassium Nitrate

NOTE: Nitric acid gives a product of greater purity.

3. Heat the mixture in a pan of simmering hot water bath until the bubbling has ceased (approximately 2 hours). The mixture will turn to a blue color.

Hot Water Bath

CAUTION: The above procedure will cause strong toxic fumes. Perform Step 3 in an open, well ventilated area.

4. Pour the hot blue solution, but not the copper, into the other pint jar. Allow solution to cool at room temperature. Crystals will form at the bottom of the jar. Discard the unreacted copper pieces in the first jar.

5. Carefully pour away the liquid from the crystals. Crush crystals into a powder with wooden rod or stick.

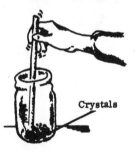

Crystals

6. Add 1/2 cup (120 milliliters) of alcohol to the powder while stirring.

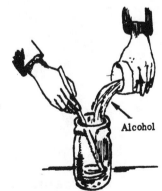

Alcohol

41

7. Filter the solution through a
paper towel into a container to collect the crystals. Wash the crystals
left on the paper towel three times,
using 1/2 cup (120 milliliters) portions of alcohol each time.

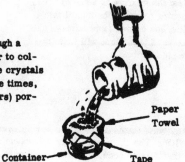

Paper
Towel

Container

Tape

8. Air dry the copper sulfate crystals for 2 hours.

NOTE: Drying time can be reduced to 1/2 hour by use of hot, not
boiling, water bath (see Step 3).

RECLAMATION OF RDX FROM C-4

RDX can be obtained from C-4 explosive with the use of gasoline. It can be used as a booster explosive for detonators (Section VI, No. 13) or as a high explosive charge.

<u>MATERIAL REQUIRED</u>:

Gasoline
C-4 explosive
2 pint glass jars, wide mouth
Paper towels
Stirring rod (glass or wood)
Water ⎫
Ceramic or glass dish ⎪
Pan   ⎬   Optional (RDX can be air dried instead)
Heat Source ⎪
Teaspoon ⎭
Cup
Tape

<u>PROCEDURE</u>:

1. Place 1-1/2 teaspoons (15 grams) of C-4 explosive in one of the pint jars. Add 1 cup (240 milliliters) of gasoline.

NOTE: These quantities can be increased to obtain more RDX. For example, use 2 gallons of gasoline per 1 cup of C-4.

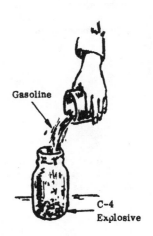

Gasoline

C-4
Explosive

2. Knead and stir the C-4 with the rod until the C-4 has broken down into small particles. Allow mixture to stand for 1/2 hour.

3. Stir the mixture again until a fine white powder remains on the bottom of the jar.

4. Filter the mixture through a paper towel into the other glass jar. Wash the particles collected on the paper towel with 1/2 cup (120 milliliters) of gasoline. Discard the waste liquid.

Tape

Dish

5. Place the RDX particles in a glass or ceramic dish. Set the dish in a pan of hot water, not boiling, and dry for a period of 1 hour.

Hot Water Bath

NOTE: The RDX particles may be air dried for a period of 2 to 3 hours.

44

TACC (TETRAMMINECOPPER (II) CHLORATE)

Tetramminecopper (II) chlorate is a primary explosive that can be made from sodium chlorate, copper sulfate and ammonia. This explosive is to be used with a booster explosive such as picric acid (Section I, No. 21) or RDX (Section I, No. 15) in the fabrication of detonators (Section 6, No. 13)

MATERIAL REQUIRED:

SOURCES:

Sodium chlorate

Section I, No. 23
Medicine
Weed killer, hardware store

Copper sulfate

Section I, No. 14
Insecticide, hardware store
Water purifying agent

Ammonia hydroxide

Household ammonia
Smelling salts

Alcohol, 95% pure
Wax, clay, pitch, etc.
Water
Bottle, narrow mouth (wine or
    coke)
Bottles, wide mouth (mason jars)
Tubing (rubber, copper, steel) to
    fit narrow mouth bottle
Teaspoon
Improvised scale   Section VII, No. 8
Heat source
Paper towel
Pan
Tape
Cup

PROCEDURE:

1. Measure 1/3 teaspoon (2-1/2 grams) of sodium chlorate into a wide mouth bottle. Add 10 teaspoons of alcohol.

45

2. Place the wide mouth bottle in a pan of hot water. Add 1 teaspoon (4 grams) of copper sulfate to the mixture. Heat for a period of 30 minutes just under the boiling point and stir occasionally.

CAUTION: Keep solution away from flame.

NOTE: Keep volume of solution constant by adding additional alcohol approximately every 10 minutes.

3. Remove solution from pan and allow to cool. Color of solution will change from blue to light green. Filter solution through a paper towel into another wide mouth bottle. Store solution until ready for step 6.

Tape

4. Add 1 cup (250 milliliters) of ammonia to the narrow mouth bottle.

5. Place tubing into the neck of bottle so that it extends about 1-1/2 inches (4 cm) inside bottle. Seal tubing to bottle with wax, clay, pitch, etc.

46

6. Place free end of tubing into the chlorate-alcohol-sulfate solution (Step 3). Heat bottle containing ammonia in a pan of hot water, but not boiling, for approximately 10 minutes.

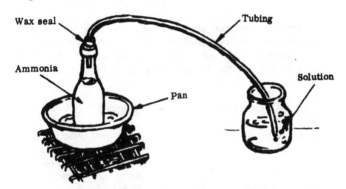

7. Bubble ammonia gas through the chlorate-alcohol-sulfate solution, approximately 10 minutes, until the color changes from light green to dark blue. Continue bubbling for another 10 minutes.

CAUTION: At this point the solution is a primary explosive. Keep away from flame.

8. Remove the solution from the pan and reduce the volume to about 1/3 of its original volume by evaporating in the open air or in a stream of air.

NOTE: Pour solution into a flat container for faster evaporation.

9. Filter the solution through a paper towel into a wide mouth bottle to collect crystals. Wash crystals with 1 teaspoon of alcohol and set aside to dry (approx. 16 hours).

CAUTION: Explosive is shock and flame sensitive. Store in a capped container.

NOTE: The drying time can be reduced to 2 hours if a hot (not boiling) water bath is used.

47

HMTD

HMTD is a primary explosive that can be made from hexamethylenetetramine, hydrogen peroxide and citric acid. This explosive is to be used with a booster explosive such as picric acid (Section I, No. 21) or RDX (Section I, No. 15) in the fabrication of detonators (Section 6, No. 13).

MATERIAL REQUIRED: SOURCES:

Hexamethylenetetramine   Drugstores under names of
                                         urotropine, hexamin,
                                         methenamine, etc.
                                     Army heat tablets.
Hydrogen peroxide   6% hair bleach (or stronger if
                                         possible)
Citric acid   Drug stores or food stores
                                         ("Sour Salt")

Containers, bottles or glasses
Paper towels
Teaspoon
Pan
Water
Tape

PROCEDURE:

1. Measure 9 teaspoons of
hydrogen peroxide into a container.

2. In 3 portions, dissolve 2-1/2
teaspoons of crushed hexamethylenetetramine in the peroxide.

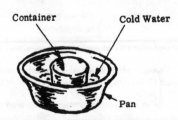

Container   Cold Water

Pan

3. Keep the solution cool for 30
minutes by placing container in a
pan of cold water.

48 4. In 5 portions, dissolve 4-1/2 teaspoons of crushed citric acid in the
hexamethylenetetramine-peroxide solution.

5. Permit solution to stand at room temperature until solid particles form at the bottom of container.

Particles

NOTE: Complete precipitation will take place in 8 to 24 hours.

CAUTION: At this point the mixture is a primary explosive. Keep away from flame.

6. Filter the mixture through a paper towel into a container to collect the solid particles.

Tape

7. Wash the solid particles collected in the paper towel with 6 teaspoons of water by pouring the water over them. Discard the liquid in the container.

8. Place these explosive particles in a container and allow to dry.

CAUTION: Handle dry explosive with great care. Do not scrape or handle it roughly. Keep away from sparks or open flames. Store in cool, dry place.

POTASSIUM OR SODIUM NITRITE AND LITHARGE (LEAD MONOXIDE)

Potassium or sodium nitrite is needed to prepare DDNP (Section I, No. 19), and litharge is required for the preparation of lead picrate (Section I, No. 20).

MATERIAL REQUIRED:	SOURCE:
Lead metal (small pieces or chips)	Plumbing supply store
Potassium (or sodium) nitrate	Field grade (Section I, No. 2)
Methyl (wood) alcohol	or Drug Store
Iron pipe with end cap	
Iron rod or screwdriver	
Paper towels	
2 glass jars, wide mouth	
Metal pan	
Heat source (hot coals or blow torch)	
Improvised scale (Section VII; No. 8)	
Cup	
Water	
Pan	

PROCEDURE:

1. Mix 12 grams of lead and 4 grams of potassium or sodium nitrate in a jar. Place the mixture in the iron pipe.

2. Heat iron pipe in a bed of hot coals or with blow torch for 30 minutes to 1 hour. (Mixture will change to a yellow color.)

50

3. Remove the iron pipe from the heat source and allow to cool. Chip out the yellow material formed in the iron pipe and place the chips in the glass jar.

Yellow Chips

4. Add 1/2 cup (120 milliliters) of methyl alcohol to the chips.

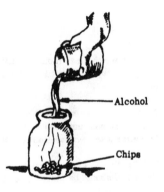

Alcohol

Chips

5. Heat the glass jar containing the mixture in a hot water bath for approximately 2 minutes (heat until there is a noticeable reaction between chips and alcohol; solution will turn darker).

Hot Water Bath

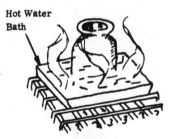

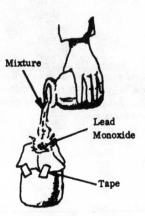

6. Filter the mixture through a
paper towel into the other glass
jar. The material left on the paper
towel is lead monoxide.

7. Remove the lead monoxide and wash it twice through a paper towel
using 1/2 cup (120 milliliters) of hot water each time. Air dry before
using.

8. Place the jar with the liquid (from Step 6) in a hot water bath (as in
Step 5) and heat until the alcohol has evaporated. The powder remaining
in the jar after evaporation is potassium or sodium nitrite.

NOTE: Nitrite has a strong tendency to absorb water from the atmos-
phere and should be stored in a closed container.

DDNP

DDNP is a primary explosive used in the fabrication of detonators (Section VI, No. 13). It is to be used with a booster explosive such as picric acid (Section I, No. 21) or RDX (Section I, No. 15).

MATERIAL REQUIRED:	SOURCES:
Picric acid	Section I, No. 21
Flowers of sulfur	
Lye (sodium hydroxide)	
Sulfuric acid, diluted	Motor vehicle batteries
Potassium or sodium nitrite	Section I, No. 18
Water	
2 glass cups, heat resistant, (Pyrex)	
Stirring rod (glass or wood)	
Improvised scale	Section VII, No. 8
Paper towels	
Teaspoon	
Tablespoon	
Eyedropper	
Heat source	
Containers	
Tape	

PROCEDURE:

1. In one of the glass cups, mix 1/2 gram of lye with 2 tablespoons (30 milliliters) of warm water.

2. Dissolve 1 teaspoon (3 grams) of picric acid in the water-lye solution. Store until ready for step 5.

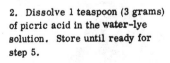

53

3. Place 1/4 teaspoon (1 milliliter) of water in the other glass cup. Add 1/2 teaspoon (2-1/2 grams) of sulfur and 1/3 teaspoon (2-1/2 grams) of lye to the water.

4. Boil solution over heat source until color turns dark red. Remove and allow solution to cool.

5. In three portions, add this sulfur-lye solution to the picric acid-lye solution (Step 2); stir while pouring. Allow mixture to cool.

6. Filter the mixture through a paper towel into a container. Small red particles will collect on the paper. Discard the liquid in the container.

Tape

7. Dissolve the red particles in 1/4 cup (60 milliliters) of boiling water.

8. Remove and filter the mixture through a paper towel as in step 6. Discard the particles left on the paper.

9. Using an eyedropper, slowly add the sulfuric acid to the filtered solution until it turns orange-brown.

10. Add 1/2 teaspoon (2-1/2 grams) more of sulfuric acid to the solution. Allow the solution to cool to room temperature.

11. In a separate container, dissolve 1/4 teaspoon (1.8 grams) of potassium or sodium nitrite in 1/3 cup (80 milliliters) of water.

12. Add this solution in one portion, while stirring, to the orange-brown solution. Allow the mixture to stand for 10 minutes. The mixture will turn light brown.

CAUTION: At this point the mixture is a primary explosive. Keep away from flame.

13. Filter the mixture through a paper towel. Wash the particles left on the paper with 4 teaspoons (20 milliliters) of water.

14. Allow the particles to dry (approx. 16 hours).

CAUTION: Explosive is shock and flame sensitive. Store explosive in a capped container.

55

NOTE: The drying time can be reduced to 2 hours if a hot (not boiling) water bath is used. See Section I, No. 16.

FOR OFFICIAL USE ONLY

PREPARATION OF LEAD PICRATE

Lead picrate is used as a primary explosive in the fabrication of detonators (Section VI, No. 13). It is to be used with a booster explosive such as picric acid (Section I, No. 21) or RDX (Section I, No. 15).

MATERIAL REQUIRED: SOURCE:

Litharge (lead monoxide)   Section I, No. 18 or plumbing
                                          supplies
Picric Acid   Section I, No. 21
Wood alcohol (methanol)   Paint removers; some antifreezes
Wooden or plastic rod
Dish or saucer (china or glass)
Teaspoon
Improvised Scale   Section VII, No. 8
Containers
Flat pan
Heat source (optional)
Water (optional)

PROCEDURE:

1. Weigh 2 grams each of picric acid and lead monoxide. Place each in a separate container.

Picric
Acid

2. Place 2 teaspoons (10 milliliters) of the alcohol in a dish. Add the picric acid to the alcohol and stir with the wooden or plastic rod.

3. Add the lead monoxide to the mixture while stirring.

CAUTION: At this point the solution is a primary explosive. Keep away from flame.

4. Continue stirring the mixture until the alcohol has evaporated. The mixture will suddenly thicken.

5. Stir mixture occasionally (to stop lumps from forming) until a powder is formed. A few lumps will remain.

CAUTION: Be very careful of dry material forming on the inside of the container.

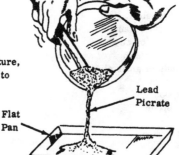

6. Spread this powdered mixture, the lead picrate, in a flat pan to air dry.

Lead Picrate

Flat Pan

NOTE: If possible, dry the mixture in a hot, not boiling, water bath for a period of 2 hours.

Hot Water Bath

PREPARATION OF PICRIC ACID FROM ASPIRIN

Picric acid can be used as a booster explosive in detonators (Section VI, No. 13), a high explosive charge, or as an intermediate to preparing lead picrate (Section I, No. 20) or DDNP (Section I, No. 19).

MATERIAL REQUIRED:

Aspirin tablets (5 grains per tablet)
Alcohol, 95% pure
Sulfuric acid, concentrated, (battery
    acid - boil until white fumes
    appear)
Potassium Nitrate (Section I, No. 2)
Water
Paper towels
Canning jar, 1 pint
Rod (glass or wood)
Glass containers
Ceramic or glass dish
Cup
Teaspoon
Tablespoon
Pan
Heat Source
Tape

PROCEDURE:

1. Crush 20 aspirin tablets in a
glass container. Add 1 teaspoon
of water and work into a paste.

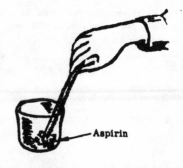

Aspirin

FOR OFFICIAL USE ONLY

2. Add approximately 1/3 to 1/2 cup of alcohol (100 milliters) to the aspirin paste; stir while pouring.

Alcohol

Aspirin Paste

3. Filter the alcohol-aspirin solution through a paper towel into another glass container. Discard the solid left on the paper towel.

Tape

4. Pour the filtered solution into a ceramic or glass dish.

Dish

5. Evaporate the alcohol and water from the solution by placing the dish into a pan of hot water. White powder will remain in the dish after evaporation.

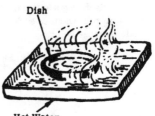

Hot Water Bath

NOTE: Water in pan should be at hot bath temperature, not boiling, approximately 160° to 180° F. It should not burn the hands.

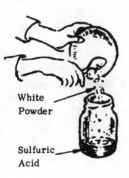

6. Pour 1/3 cup (80 milliliters) of concentrated sulfuric acid into a canning jar. Add the white powder to the sulfuric acid.

White Powder

Sulfuric Acid

Hot Water Bath

7. Heat canning jar of sulfuric acid in a pan of simmering hot water bath for 15 minutes; then remove jar from the bath. Solution will turn to a yellow-orange color.

8. Add 3 level teaspoons (15 grams) of potassium nitrate in three portions to the yellow-orange solution; stir vigorously during additions. Solution will turn red, and then back to a yellow-orange color.

Solution

9. Allow the solution to cool to ambient or room temperature while stirring occasionally.

60

10. Slowly pour the solution, while stirring, into 1-1/4 cup (300 milliliters) of cold water and allow to cool.

Cold Water

11. Filter the solution through a paper towel into a glass container. Light yellow particles will collect on the paper towel.

12. Wash the light yellow particles with 2 tablespoons (25 milliliters) of water. Discard the waste liquid in the container.

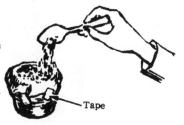

Tape

13. Place particles in ceramic dish and set in a hot water bath, as in step 5, for 2 hours.

DOUBLE SALTS

Double Salts is used as a primary explosive in the fabrication of detonators (Section VI, No. 13). It can be made in the field from silver (coins), nitric acid, calcium carbide, and water.

<u>MATERIALS REQUIRED</u>:

Nitric acid (90% conc.) (Section I, No. 4)
Silver metal (silver coin, about 5/8 in diameter)
Calcium carbide (acetylene or calcium carbide lamps)
Rubber and glass tubing (approx. 1/4 in. inside diameter)
Paper towels
Heat-resistant bottles or ceramic jugs, 1 to 2 quart
    capacity, and one cork to fit. (Punch hole in cork
    to fit tubing.)
Teaspoon (aluminum, stainless steel or wax-coated) or equivalent
    measure
Glass container
Heat source
Long narrow jar (olive jar)
Tape
Water
Alcohol

<u>PROCEDURE</u>:

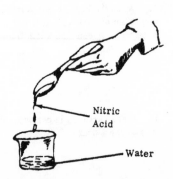

1. Dilute 2-1/4 teaspoons of nitric acid with 1-1/2 teaspoons of water in a glass container by adding the acid to the water.

2. Dissolve a silver coin (a silver dime) in the diluted nitric acid. The solution will turn to a green color.

NOTE: It may be necessary to warm the container to completely dissolve the silver coin.

FOR OFFICIAL USE ONLY

CAUTION: Acid will burn skin and destroy clothing. If any is spilled, wash it away with a large quantity of water. Do not inhale fumes.

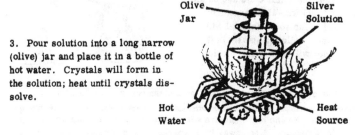

3. Pour solution into a long narrow (olive) jar and place it in a bottle of hot water. Crystals will form in the solution; heat until crystals dissolve.

4. While still heating and after crystals have dissolved, place 10 teaspoons of calcium carbide in another glass bottle and add 1 teaspoon of water. After the reaction has started add another teaspoon of water. Then set up as shown.

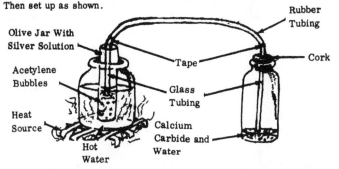

5. Bubble acetylene through the solution for 5 to 8 minutes. A brown vapor will be given off and white flakes will appear in the silver solution.

6. Remove the silver solution from the heat source and allow it to cool. Filter the solution through a paper towel into a glass container. Green crystals will collect on the paper.

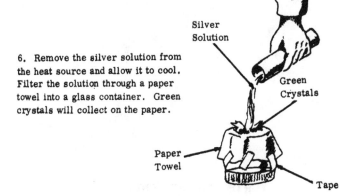

63

7. Wash the solids collected on the paper towel with 12 teaspoons of alcohol. The solid material will turn white while the solvent in the container will have a green color.

White Solids

Green Solvent

8. Place the white solid material on a clean paper towel to air dry.

CAUTION: Handle dry explosive with great care. Do not scrape or handle it roughly. Keep away from sparks or open flames. Store in cool, dry place.

SODIUM CHLORATE

Sodium chlorate is a strong oxidizer used in the manufacture of explosives. It can be used in place of potassium chlorate (see Section I, No. 1).

MATERIAL REQUIRED:

2 carbon or lead rods (1 in.
    diameter x 5 in. long)

Salt or, ocean water
Sulfuric acid, diluted
Motor vehicle
Water
2 wires, 16 gauge (3/64 in.
    diameter approx.), 6 ft. long,
    insulated
Gasoline
1 gallon glass jar, wide mouth
    (5 in. diameter x 6 in. high
    approx.)
Sticks
String
Teaspoon
Trays
Cup
Heavy cloth
Knife
Large flat pan or tray

SOURCES:

Dry cell batteries (2-1/2 in.
    diameter x 7 in. long) or
    plumbing supply store
Grocery store or ocean
Motor vehicle batteries

PROCEDURE:

1. Mix 1/2 cup of salt into the one gallon glass jar with 3 liters (3 quarts) of water.

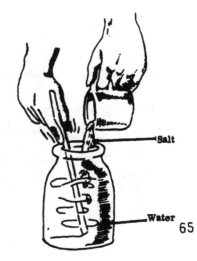

Salt

Water

65

2.   Add 2 teaspoons of battery acid to the solution and stir vigorously for 5 minutes.

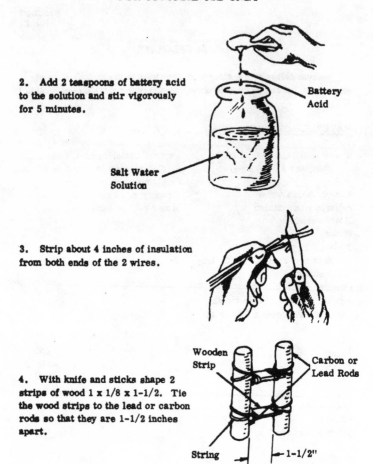

Battery Acid

Salt Water Solution

3.   Strip about 4 inches of insulation from both ends of the 2 wires.

4.   With knife and sticks shape 2 strips of wood 1 x 1/8 x 1-1/2. Tie the wood strips to the lead or carbon rods so that they are 1-1/2 inches apart.

Wooden Strip

Carbon or Lead Rods

String

1-1/2"

5.   Connect the rods to the battery in a motor vehicle with the insulated wire.

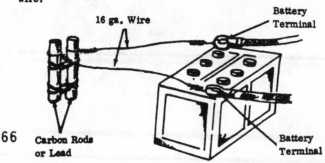

16 ga. Wire

Battery Terminal

66   Carbon Rods or Lead

Battery Terminal

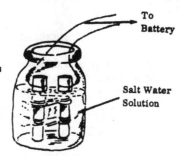

To
Battery

Salt Water
Solution

6. Submerge 4-1/2 in. of the rods
into the salt water solution.

7. With gear in neutral position, start the vehicle engine. Depress the
accelerator approximately 1/5 of its full travel.

8. Run the engine with the accelerator in this position for 2 hours; then,
shut it down 2 hours.

9. Repeat this cycle for a total of 64 hours while maintaining the level
of the acid-salt water solution in the glass jar.

CAUTION: This arrangement employs voltages which may be dangerous
to personnel. Do not touch bare wire leads while engine is running.

10. Shut off the engine. Remove the rods from the glass jar and dis-
connect wire leads from the battery.

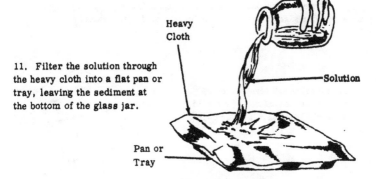

Heavy
Cloth

Solution

11. Filter the solution through
the heavy cloth into a flat pan or
tray, leaving the sediment at
the bottom of the glass jar.

Pan or
Tray

12. Allow the water in the filtered solution to evaporate at room temperature (approx. 16 hours). The residue is approximately 60% or
more sodium chlorate which is pure enough to be used as an explosive
ingredient.

67

MERCURY FULMINATE

Mercury Fulminate is used as a primary explosive in the fabrication of detonators (Section VI, No. 13). It is to be used with a booster explosive such as picric acid (Section I, No. 21) or RDX (Section I, No. 15).

MATERIAL REQUIRED:	SOURCE:
Nitric Acid, 90% conc. (1.48 sp. gr.)	Field grade (Section I, No. 4) or industrial metal processors
Mercury	Thermometers, mercury switches, old radio tubes
Ethyl (grain) alcohol (90%)	
Filtering material	Paper towels
Teaspoon measure (1/4, 1/2, and 1 teaspoon capacity) - aluminum, stainless steel or wax-coated	
Heat source	
Clean wooden stick	
Clean water	
Glass containers	
Tape	
Syringe	

PROCEDURE:

1. Dilute 5 teaspoons of nitric acid with 2-1/2 teaspoons of clean water in a glass container by adding the acid to the water.

2. Dissolve 1/8 teaspoon of mercury in the diluted nitric acid. This will yield dark red fumes.

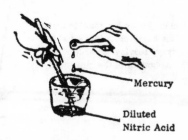

Mercury

Diluted
Nitric Acid

NOTE: It may be necessary to add water, one drop at a time, to the mercury-acid solution in order to start reaction.

CAUTION: Acid will burn skin and destroy clothing. If any is spilled, wash it away with a large quantity of water. Do not inhale fumes.

3. Warm 10 teaspoons of the alcohol in a container until the alcohol feels warm to the inside of the wrist.

Alcohol

Heat Source

4. Pour the metal-acid solution into the warm alcohol. Reaction should start in less than 5 minutes. Dense white fumes will be given off during reaction. As time lapses, the fumes will become less dense. Allow 10 to 15 minutes to complete reaction. Fulminate will settle to bottom.

Metal-Acid Solution

Warm Alcohol

CAUTION: This reaction generates large quantities of toxic, flammable fumes. The process must be conducted outdoors or in a well ventilated area, away from sparks or open flames. Do not inhale fumes.

5. Filter the solution through a paper towel into a container. Crystals may stick to the side of the container. If so, tilt and squirt water down the sides of the container until all the material collects on the filter paper.

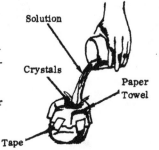

Solution

Crystals

Paper Towel

Tape

69

6. Wash the crystals with 6 tea-spoons of ethyl alcohol.

7. Allow these mercury fulminate crystals to air dry.

CAUTION: Handle dry explosive with great care. Do not scrape or handle it roughly. Keep away from sparks or open flames. Store in cool, dry place.

SODIUM CHLORATE AND SUGAR OR ALUMINUM EXPLOSIVE

An explosive munition can be made from sodium chlorate combined with granular sugar, or aluminum powder. This explosive can be detonated with a No. 8 commercial or a Military J-2 blasting cap.

MATERIAL REQUIRED: SOURCE:

Sodium chlorate   Section I, No. 23
Granular sugar   Food store
Aluminum powder   Paint store
Wooden rod or stick
Bottle or jar
Blasting cap
Steel pipe (threaded at one end), end cap
   and tape
Wax
Measuring container (cup, quart, etc.)

PROCEDURE:

1. Add three volumes (cups, quarts, etc.) sodium chlorate to one volume aluminum, or two granular sugar, in bottle or jar.

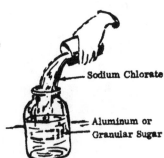

Sodium Chlorate

Aluminum or Granular Sugar

2. Mix ingredients well by stirring with the wooden rod or stick.

HOW TO USE:

1. Wax blasting cap, pipe and end cap.

2. Thread end cap onto pipe.

3. Pour mixture into pipe.

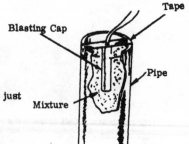

4. Insert and tape blasting cap just beneath surface of mixture.

72 NOTE: Confining the open end of the pipe will add to the effectiveness of the explosive.

PIPE HAND GRENADE

Hand grenades can be made from a piece of iron pipe. The filler can be plastic or granular military explosive, improvised explosive, or propellant from shotgun or small arms ammunition.

MATERIAL REQUIRED

Iron pipe, threaded ends, 1 1/2"
   to 3" diam., 3" to 8" long.
Two (2) iron pipe caps.
Explosive or propellant
Nonelectric blasting cap.
   (Commercial or military)
Fuse cord
Hand drill
Pliers

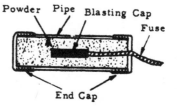

PROCEDURE

1. Place blasting cap on one end of fuse cord and crimp with pliers.

   NOTE: To find out how long the fuse cord should be, check the time it takes a known length to burn. If 12 inches burns in 30 seconds, a 6-inch cord will ignite the grenade in 15 seconds.

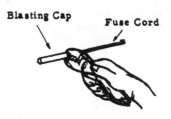

2. Screw pipe cap to one end of pipe. Place fuse cord with blasting cap into the opposite end so that the blasting cap is near the center of the pipe.

   NOTE: If plastic explosive is to be used, fill pipe before inserting blasting cap. Push a round stick into the center of the explosive to make a hole and then insert the blasting cap.

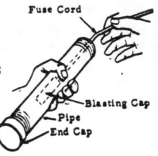

73

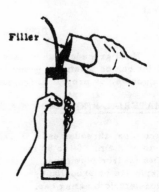

Filler

3. Pour explosive or propellant into pipe a little bit at a time. Tap the base of the pipe frequently to settle filler.

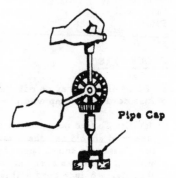

Pipe Cap

4. Drill a hole in the center of the unassembled pipe cap large enough for the fuse cord to pass through.

5. Wipe pipe threads to remove any filler material.

Slide the drilled pipe cap over the fuse and screw handtight onto the pipe.

74

NAIL GRENADE

Effective fragmentation grenades can be made from a block of TNT or other blasting explosive and nails

MATERIAL REQUIRED:

Block of TNT or other blasting
 explosive
Nails
Non-Electric Military blasting cap
Fuse Cord
Tape, string, wire or glue

PROCEDURE:

1. If an explosive charge other
than a standard TNT block is
used, make a hole in the center
of the charge for inserting the
blasting cap. TNT can be drilled
with relative safety. With
plastic explosives, a hole can
be made by pressing a round
stick into the center of the charge.
The hole should be deep enough
that the blasting cap is totally
within the explosive.

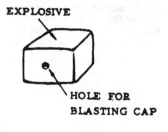

EXPLOSIVE

HOLE FOR
BLASTING CAP

2. Tape, tie or glue one or
two rows of closely packed nails
to sides of explosive block.
Nails should completely cover
the four surfaces of the block.

TAPE

EXPLOSIVE

NAILS

3. Place blasting cap on one
end of the fuse cord and crimp
with pliers.

BLASTING CAP FUSE CORD

NOTE: To find out how long the
fuse cord should be, check the
time it takes a known length
to burn. If 12 inches (30 cm)
burns for 30 seconds, a 10
second delay will require a 4
inch (10 cm) fuse.

75

4. Insert the blasting cap in the hole in the block of explosive. Tape or tie fuse cord securely in place so that it will not fall out when the grenade is thrown.

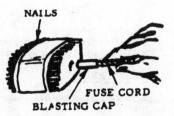

NAILS

FUSE CORD

BLASTING CAP

ALTERNATE USE:

An effective directional anti-personnel mine can be made by placing nails on only one side of the explosive block. For this case, an electric blasting cap can be used.

TARGET

76

WINE BOTTLE CONE CHARGE

This cone charge will penetrate 3 to 4 inches of armor.
Placed on an engine or engine compartment it will disable a tank
or other vehicle.

MATERIAL REQUIRED:

Glass wine bottle with false bottom (cone shaped).
Plastic or castable explosive
Blasting cap
Gasoline or Kerosene (small amount)
String
Adhesive tape

PROCEDURE:

1. Soak a piece of string in gasoline or kerosene. Double wrap
this string around the wine bottle
approximately 3 in. (7 1/2 cm)
above the top of the cone.

NOTE: A small amount of motor
oil added to the gasoline or
kerosene will improve results.

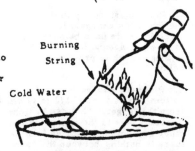

2. Ignite the string and allow to
burn for 1 to 2 minutes. Then
plunge the bottle into cold water
to crack the bottle. The top
half can now be easily removed
and discarded.

3. If plastic explosive is used:
(a) pack explosive into the bottle
a little at a time compressing
with a wooden rod. Fill the
bottle to the top.

(b) press a 1/4 in. wooden dowel
1/2 in. (12mm) into the middle of
the top of the explosive charge to
form a hole for the blasting cap.

4. If TNT or other castable explosive is used:
(a) break explosive into small pieces using a wooden mallet or
non-sparking metal tools. Place pieces in a tin can.

77

(b) Suspend this can in a larger container which is partly filled with water. A stiff wire or stick pushed through the smaller can will accomplish this.

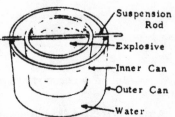

Suspension Rod
Explosive
Inner Can
Outer Can
Water

CAUTION: The inner can must not rest on the bottom of the outer container.

(c) Heat the container on an electric hot plate or other heat source. Stir the explosive frequently with a wooden stick while it is melting.

CAUTION: Keep area well ventilated while melting explosive. Fumes may be poisonous.

(d) When all the explosive has melted, remove the inner container and stir the molten explosive until it begins to thicken. During this time the bottom half of the wine bottle should be placed in the container of hot water. This will pre-heat the bottle so that it will not crack when the explosive is poured.

(e) Remove the bottle from hot water and dry thoroughly. Pour molten explosive into the bottle and allow to cool. The crust which forms on top of the charge during cooling should be broken with a wooden stick and more explosive added. Do this as often as necessary until the bottle is filled to the top.

(f) When explosive has completely hardened, bore a hole for the blasting cap in the middle of the top of the charge about 1/2 in. (12mm) deep.

HOW TO USE:

1. Place blasting cap in the hole in the top of the charge. If non-electric cap is used be sure cap is crimped around fuze and fuze is long enough to provide safe delay.

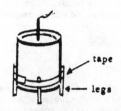

2. Place the charge so that the bottom is 3 to 4 in. (7 1/2 to 10 cm) from the target. This can be done by taping legs to the charge or any other convenient means as long as there is nothing between the base of the charge and the target.

tape
legs

3. If electric cap is used, connect blasting cap wires to firing circuit.

Sand or Dirt

Container

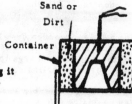

NOTE: The effectiveness of this charge can be increased by placing it inside a can, box, or similar container and packing sand or dirt between the charge and the container.

78

GRENADE-TIN CAN LAND MINE

This device can be used as a land mine that will explode when the
trip wire is pulled.

MATERIAL REQUIRED:

Hand grenade having side safety lever
Sturdy container, open at one end, that is just large enough to fit over
   grenade and its safety lever (tin can of proper size is suitable).
Strong string or wire

NOTE: The container must be of such a size that, when the grenade is
placed in it and the safety pin removed, its sides will prevent the safety
lever from springing open. One end must be completely open.

PROCEDURE:

1.   Fasten one piece of string to
the closed end of container, making
a strong connection. This can be
done by punching 2 holes in the can,
looping the string through them, and
tying a knot.

2.   Tie free end of this string to bush, stake, fencepost, etc.

3.   Fasten another length of string
to the grenade such that it cannot
interfere with the functioning of the
ignition mechanism of the grenade.

4.   Insert grenade into container.

79

5. Lay free length of string across
path and fasten to stake, bush, etc.
The string should remain taut.

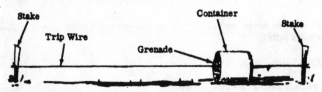

HOW TO USE:

1. Carefully withdraw safety pin by pulling on ring. Be sure safety
lever is restrained during this operation. Grenade will function in
normal manner when trip wire is pulled.

NOTE: In areas where concealment is possible, a greater effect may
be obtained by suspending the grenade several feet above ground, as
illustrated below.

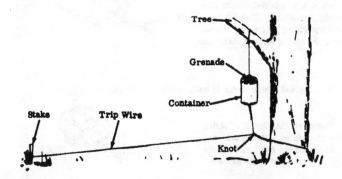

80

MORTAR SCRAP MINE

A directional shrapnel launcher that can be placed in the path of advancing troops.

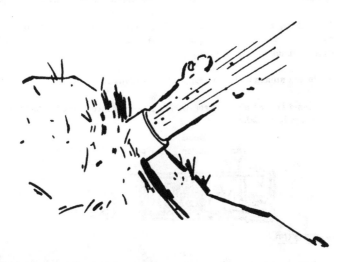

MATERIAL REQUIRED:

Iron pipe approximately 3 ft. (1 meter) long and 2 in. to 4 in. (5 to
   10 cm) in diameter and threaded on at least one end. Salvaged
   artillery cartridge case may also be used.
Threaded cap to fit pipe.
Black powder or salvaged artillery propellant about 1/2 lb. (200 gms)
   total.
Electrical igniter (commercial SQUIB or improvised igniter, Section
   VI, No. 1). Safety or improvised fuse may also be used.
Small stones about 1 in. (2-1/2 cm) in diameter or small size scrap;
   about 1 lb. (400 gms) total.
Rags for wadding, each about 20 in. by 20 in. (50 cm x 50 cm)
Paper or bag
Battery and wire
Stick (non-metallic)

Note: Be sure pipe has no cracks or flaws.

PROCEDURE:

1. Screw threaded cap onto pipe.

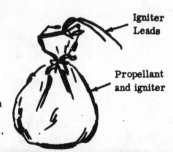

Igniter Leads

Propellant and igniter

2. Place propellant and igniter in paper or rag and tie package with string so contents will not fall out.

3. Insert packaged propellant and igniter into pipe until package rests against threaded cap leaving firing leads extending from open end of pipe.

4. Roll rag till it is about 6 in. (15-1/2 cm) long and the same diameter as pipe. Insert rag wadding against packaged propellant igniter. With caution, pack tightly using stick.

5. Insert stones and/or scrap metal into pipe.

6. Insert second piece of rag wadding against stones and/or metal scrap. Pack tightly as before.

Wad   Metal Scrap   Wad   Firing Leads

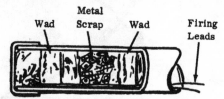

HOW TO USE:

1. Bury pipe in ground with open end facing the expected path of the enemy. The open end may be covered with cardboard and a thin layer of dirt or leaves as camouflage.

2. Connect firing leads to battery and switch. Mine can be remotely fired when needed or attached to trip device placed in path of advancing troops.

NOTE: A NON-ELECTRICAL ignition system can be substituted for the electrical ignition system as follows.

1. Follow above procedure, substituting safety fuse for igniter.

2. Light safety fuse when ready to fire.

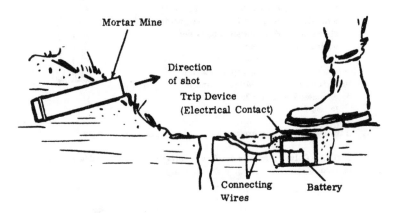

Mortar Mine

Direction of shot

Trip Device (Electrical Contact)

Connecting Wires

Battery

COKE BOTTLE SHAPED CHARGE

This shaped charge will penetrate 3 in. (7-1/2 cm) of armor. (It will disable a vehicle if placed on the engine or engine compartment).

MATERIAL REQUIRED:

Glass Coke bottle, 6-1/2 oz. size
Plastic or castable explosive, about
   1 lb. (454 gm)
Blasting cap
Metal cylinder, open at both ends, about
   6 in. (15 cm) long and 2 in. (5 cm) inside
   diameter. Cylinder should be heavy
   walled for best results.
Plug to fit mouth of coke bottle
   (rags, metal, wood, paper, etc.)
Non-metal rod about 1/4 in. (6 mm) in
   diameter and 8 in. (20 cm) or more
   in length.
Tape or string
2 tin cans if castable explosive is used (See Section II, No. 3)

Coke
Bottle

NOTE: Cylinder may be cardboard, plastic, etc. if castable explosive is used.

PROCEDURE:

Plug

1.   Place plug in mouth of bottle.

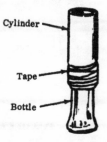

Cylinder

Tape

Bottle

2.   Place cylinder over top of bottle until bottom of cylinder rests on widest part of bottle. Tape cylinder to bottle. Container should be straight on top of bottle.

84

3. If plastic explosive is used:

a. Place explosive in cylinder a little at a time tamping with rod until cylinder is full.

b. Press the rod about 1/2 in. (1 cm) into the middle of the top of the explosive charge to form a hole for the blasting cap.

4. If castable explosive is used, follow procedure of Wine Bottle Cone Charge, Section II, No. 3, Step 4, a through f.

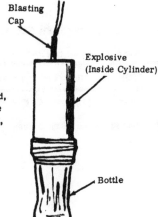

Blasting Cap

Explosive (Inside Cylinder)

Bottle

HOW TO USE:

Method 1. If electrical blasting cap is used:

1. Place blasting cap in hole in top of explosive.

CAUTION: Do not insert blasting cap until charge is ready to be detonated.

85

2. Place bottom of Coke Bottle flush against the target. If target is not flat and horizontal, fasten bottle to target by any convenient means, such as by placing tape or string around target and top of bottle. Bottom of bottle acts as stand-off.

CAUTION: Be sure that base of bottle is flush against target and that there is nothing between the target and the base of the bottle.

3. Connect leads from blasting cap to firing circuit.

Method II: If non-electrical blasting cap is used:

1. Crimp cap around fuse.

CAUTION: Be sure fuse is long enough to provide a safe delay.

2. Follow steps 1, 2, and CAUTIONS of Method I.

3. Light fuse when ready to fire.

CYLINDRICAL CAVITY SHAPED CHARGE

A shaped charge can be made from common pipe. It will penetrate 1-1.2 in. (3-1/2 cm) of steel, producing a hole 1-1/2 in. (3-1/2 cm) in diameter.

MATERIAL REQUIRED:

Iron or steel pipe, 2 to 2-1/2 in. (5 to 6-1/2 cm) in diameter and 3 to 4 in. (7-1/2 to 10 cm) long
Metal pipe, 1/2 to 3/4 in. (1-1/2 to 2 cm) in diameter and 1-1/2 in. (3-1/2 cm) long, open at both ends. (The wall of the pipe should be as thin as possible.)
Blasting cap
Non-metallic rod, 1/4 in. (6 mm) in diameter
Plastic or castable explosive
2 metal cans of different sizes ⎫
Stick or wire   ⎬   If castable explosive is used
Heat source   ⎭

PROCEDURE:

1. If plastic explosive is used:

   a. Place larger pipe on flat surface. **Hand** pack and tamp explosive into pipe. Leave approximately 1/4 in. (6 mm) space at top.

   Approximately 1/4 in. Empty Space
   Large Pipe
   Plastic Explosive
   Flat Surface

   b. Push rod into **center** of explosive. Enlarge hole in explosive to diameter and length of small pipe.

   1-1/2 in.

   c. Insert small pipe into hole.

   1/4 in. Empty Space
   Small Pipe
   Large Pipe

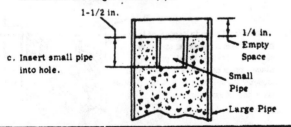

IMPORTANT: Be sure direct contact is made between explosive and small pipe. Tamp explosive around pipe **by hand** if necessary.

d. Make sure that there is 1/4 in. (6 mm) empty space above small pipe. Remove explosive if necessary.

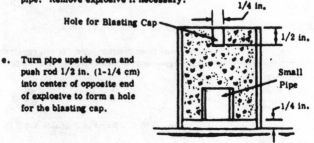

Hole for Blasting Cap

1/4 in.

1/2 in.

Small Pipe

1/4 in.

e. Turn pipe upside down and push rod 1/2 in. (1-1/4 cm) into center of opposite end of explosive to form a hole for the blasting cap.

CAUTION: Do not insert blasting cap in hole until ready to fire shaped charge.

---

2. If TNT or other castable explosive is used:

a. Follow procedure, Section II, No. 3, Step 4, Parts a, b, c, including CAUTIONS.

b. When all the explosive has melted, remove the inner container and stir the molten explosive until it begins to thicken.

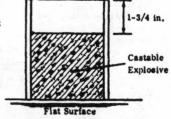

1-3/4 in.

Castable Explosive

c. Place large pipe on flat surface. Pour explosive into pipe until it is 1-3/4 in. (4 cm) from the top.

Flat Surface

Small Pipe

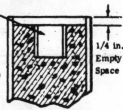

1/4 in. Empty Space

d. Place small pipe in <u>center</u> of large pipe so that it rests on top of explosive. Holding small pipe in place, pour explosive around small pipe until explosive is 1/4 in. (6 mm) from top of large pipe.

e. Allow explosive to cool. Break crust that forms on top of the charge during cooling with a wooden stick and add more explosive. Do this as often as necessary until explosive is 1/4 in. (6 mm) from top.

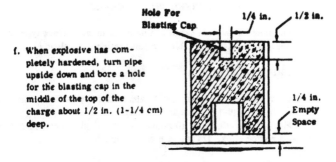

f. When explosive has completely hardened, turn pipe upside down and bore a hole for the blasting cap in the middle of the top of the charge about 1/2 in. (1-1/4 cm) deep.

HOW TO USE:

Method I - If electrical blasting cap is used:

1. Place blasting cap in hole made for it.

> CAUTION: Do not insert blasting cap until charge is ready to fire.

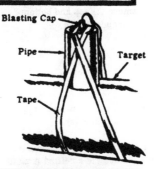

2. Place other end of pipe flush against the target. Fasten pipe to target by any convenient means, such as by placing tape or string around target and top of pipe, if target is not flat and horizontal.

> CAUTION: Be sure that base of pipe is flush against target and that there is nothing between the target and the base of the pipe.

3. Connect leads from blasting cap to firing circuit.

Method II - If non-electrical blasting cap is used:

1. Crimp cap around fuse.

> CAUTION: Be sure fuse is long enough to provide a safe delay.

2. Follow Steps 1, 2, and CAUTION of Method I.

3. Light fuse when ready to fire.

FUNNEL SHAPED CHARGE

An effective shaped charge can be made using various types of commercial funnels. See table for penetration capabilities.

MATERIAL REQUIRED:

Container (soda or beer can, etc.), approximately 2-1/2 in. diameter
    x 5 in. long (6-1/4 cm x 12-1/2 cm)
Funnel(s) (glass, steel, or aluminum) 2-1/2 in. (6-1/2 cm) in diameter
Wooden rod or stick, 1/4 in. (6 mm) in diameter
Tape
Blasting cap (electrical or non-electrical)
Sharp cutting edge
Explosive

PROCEDURE:

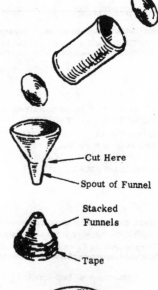

1. Remove the top and bottom from can and discard.

2. Cut off and throw away the spout of the funnel(s).

Cut Here

Spout of Funnel

NOTE: When using 3 funnels (see table), place the modified funnels together as tight and as straight as possible. Tape the funnels together at the outer ridges.

Stacked
Funnels

Tape

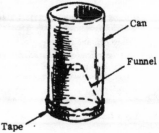

Can

Funnel

3. Place the funnel(s) in the modified can. Tape on outer ridges to hold funnel(s) to can.

Tape

90

4. If plastic explosive is used,
fill the can with the explosive
using small quantities, and tamp
with wooden rod or stick.

Explosive

NOTE: If castable explosive is
used, refer to step 4 of Section II,
No. 3.

5. Cut wooden rod to lengths 3 inches
longer than the standoff length.
(See table.) Position three of these
rods around the explosive filled can
and hold in place with tape.

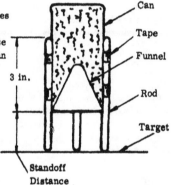

Can

Tape

Funnel

3 in.

Rod

NOTE: The position of the rods
on the container must conform to
standoff dimensions to obtain the
penetrations given in the table.

Target

Standoff
Distance

Table

Funnel Material	No. of Funnels	Standoff		Penetration	
		inches	metric	inches	metric
Glass	1	3-1/2	9 cm	4	10 cm
Steel	3	1	2-1/2 cm	2-1/2	6 cm
Aluminum	3	3-1/2	9 cm	2-1/2	6 cm
*If only one steel or aluminum funnel is available:					
Steel	1	1	2-1/2 cm	1-1/2	4 cm
Aluminum	1	1	2-1/2 cm	1-1/2	4 cm

6. Make a hole for blasting cap in the center of the explosive with rod or stick.

CAUTION: Do not place blasting cap in place until the Funnel Shaped Charge is ready for use.

HOW TO USE:

1. Place blasting cap in the hole in top of the charge. If non-electric cap is used, be sure cap is crimped around fuse and fuse is long enough to provide safe delay.

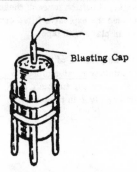

Blasting Cap

2. Place (tape if necessary) the Funnel Shaped Charge on the target so so that nothing is between the base of charge and target.

3. If electric cap is used, connect blasting cap wires to firing circuit.

LINEAR SHAPED CHARGE

This shaped charge made from construction materials will cut through up to nearly 3 inches of armor depending upon the liner used (see table).

MATERIAL REQUIRED:

Standard structural angle or pipe (see table)
Wood or cardboard container
Hacksaw ⎫
Vice ⎭ if pipe is used
Wooden rod, 1/4 in. (6 mm) diameter
Explosive
Blasting cap
Tape

Table

Type	Material	Liner Size in. - Nom.	Standoff in.	Standoff metric	Penetration in.	Penetration metric
angle	steel	3 x 3 legs x 1/4 web	2	5 cm	2-3/4	7 cm
angle	aluminum	2 x 2 legs x 3/16 web	5-1/2	14 cm	2-1/2	6 cm
pipe half section	aluminum	2 diameter	2	5 cm	2	5 cm
pipe half section	copper	2 diameter	1	2-1/2 cm	1-3/4	4 cm

NOTE: These were the only linear shaped charges of this type that were found to be more efficient than the Ribbon Charge.

Ribbon Charge: No standoff is required; just place on target.

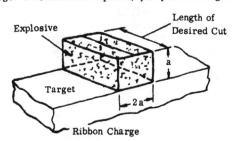

93

PROCEDURE:

1. If pipe is used --

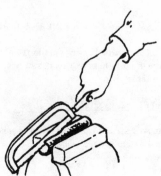

   a. Place the pipe in the vise
      and cut pipe in half length-
      wise. Remove the pipe
      half sections from vise.

   b. Discard one of the pipe
      half sections, or save for
      another charge.

2. Place angle or pipe half section with open end face down on a flat surface.

3. Make container from any material available. The container must be as wide as the angle or pipe half section, twice as high, and as long as the desired cut to be made with the charge.

4. Place container over the liner (angle or pipe half section) and tape liner to container.

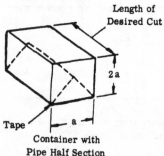

Length of
Desired Cut

2 a

Tape ⊢— a —⊣

Container with
Pipe Half Section

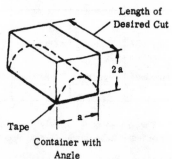

Length of
Desired Cut

2 a

Tape ⊢— a —⊣

Container with
Angle

5. If plastic explosove is used, fill
the container with the explosive
using small quantities, and tamp
with wooden rod or stick.

NOTE: If castable explosive is
used, refer to step 4 of Section II,
No. 3.

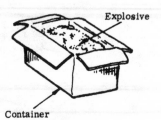

Explosive

Container

94

6. Cut wooden rod to lengths 2 inches longer than the standoff length (see table). Position the rods at the corners of the explosive filled container and hold in place with tape.

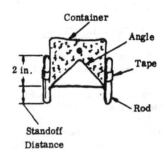

NOTE: The position of the rods on the container must conform to standoff and penetration dimensions given in the table.

7. Make a hole for blasting cap in the side of the container 1/2 in. above the liner and centered with the wooden rod.

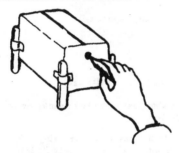

CAUTION: Do not place blasting cap in place until the Linear Shaped Charge is ready for use.

HOW TO USE:

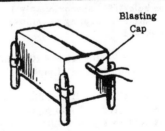

1. Place blasting cap into hole on the side of the container. If non-electric cap is used, be sure cap is crimped around fuse and fuse is long enough to provide safe delay.

2. Place (tape if necessary) the Linear Shaped Charge on the target so that nothing is between base of charge and target.

3. If electric cap is used, connect blasting cap wires to firing circuit.

95

PIPE PISTOL FOR 9 MM AMMUNITION

A 9 mm pistol can be made from 1/4" steel gas or water pipe and fittings.

MATERIAL REQUIRED

1/4" nominal size steel pipe 4 to 6
   inches long with threaded ends.
1/4" Solid pipe plug
Two (2) steel pipe couplings
Metal strap - roughly 1/8" x
   1/4" x 5"
Two (2) elastic bands
Flat head nail - 6D or 8D (approx
   1/16" diameter)
Two (2) wood screws #8
Wood 8" x 5" x 1"
Drill
1/4" wood or metal rod, (approx
   8" long)

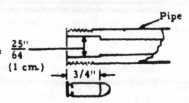

PROCEDURE

1. Carefully inspect pipe and fittings.

    a. Make sure that there are NO cracks or other flaws in the
       pipe or fittings.

    b. Check inside diameter of pipe using a 9 mm cartridge as a
       gauge. The bullet should closely fit into the pipe without for-
       cing but the cartridge case SHOULD NOT fit into pipe.

    c. Outside diameter of pipe MUST NOT BE less than 1 1/2
       times bullet diameter (.536 inches; 1.37 cm)

2. Drill a 9/16" (1.43 cm) diameter hole 3/8" (approximately 1
cm) into one coupling to remove
the thread.

   Drilled section should fit tightly
over smooth section of pipe.

3. Drill a 25/64" (1 cm) diameter
hole 3/4" (1.9 cm) into pipe. Use
cartridge as a gauge; when a cartridge is inserted into the pipe, the
base of the case should be even
with the end of the pipe. Thread
coupling tightly onto pipe, drilled
end first.

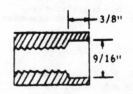

4. Drill a hole in the center of the pipe plug just large enough for the nail to fit through.

   HOLE MUST BE CENTERED IN PLUG.

5. Push nail through plug until head of nail is flush with square end. Cut nail off at other end 1/16" (.158 cm) away from plug. Round off end of nail with file.

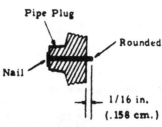

Pipe Plug
Rounded
Nail
1/16 in.
(.158 cm.)

6. Bend metal strap to "U" shape and drill holes for wood screws. File two small notches at top.

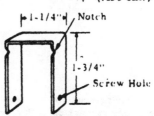

1-1/4" Notch
1-3/4"
Screw Hole

7. Saw or otherwise shape 1" (2.54 cm) thick hard wood into stock.

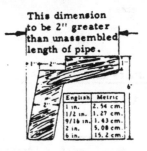

This dimension to be 2" greater than unassembled length of pipe.

English	Metric
1 in.	2.54 cm.
1/2 in.	1.27 cm.
9/16 in.	1.43 cm.
2 in.	5.08 cm.
6 in.	15.2 cm.

8. Drill a 9/16" diameter (1.43 cm) hole through the stock. The center of the hole should be approximately 1/2" (1.27 cm) from the top.

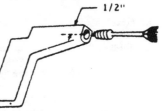

1/2"

9. Slide the pipe through this hole and attach front coupling. Screw drilled plug into rear coupling.

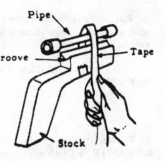

NOTE: If 9/16" drill is not available cut a "V" groove in the top of the stock and tape pipe securely in place.

10. Position metal strap on stock so that top will hit the head of the nail. Attach to stock with wood screw on each side.

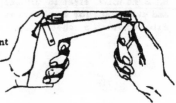

11. String elastic bands from front coupling to notch on each side of the strap.

SAFETY CHECK - TEST FIRE PISTOL BEFORE HAND FIRING

1. Locate a barrier such as a stone wall or large tree which you can stand behind in case the pistol ruptures when fired.

2. Mount pistol solidly to a table or other rigid support at least ten feet in front of the barrier.

3. Attach a cord to the firing strap on the pistol.

4. Holding the other end of the cord, go behind the barrier.

5. Pull the cord so that the firing strap is held back.

6. Release the cord to fire the pistol. (If pistol does not fire, shorten the elastic bands or increase their number.)

IMPORTANT: Fire at least five rounds from behind the barrier and then re-inspect the pistol before you attempt to hand fire it.

98

HOW TO OPERATE PISTOL

1. To Load

   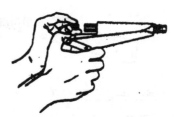

   a. Remove plug from rear coupling.

   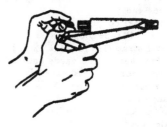

   b. Place cartridge into pipe.

   c. Replace plug.

2. To Fire

   a. Pull strap back and hold with thumb until ready.

   b. Release strap.

3. To Remove Shell Case

   a. Remove plug from rear coupling.

   b. Insert 1/4" diameter steel or wooden rod into front of pistol and push shell case out.

SHOTGUN (12 GAUGE)

A 12-gauge shotgun can be made from 3/4" water or gas pipe and fittings.

MATERIALS REQUIRED

Wood 2" x 4" x 32"
3/4" nominal size water or gas pipe 20" to 30" long threaded on
 one end.
3/4" steel coupling
Solid 3/4" pipe plug
Metal strap (1/4" x 1/16" x 4")
Twine, heavy (100 yards approximately)
3 wood screws   and screwdriver
Flat head nail 6D or 8D
Hand drill
Saw or knife
File
Shellac or lacquer
Elastic Bands
PROCEDURE

1. Carefully inspect pipe and fittings.

    a. Make sure that there are no cracks or other flaws.

    b. Check inside diameter of pipe. A 12-gauge shot shell should
       fit into the pipe but the brass rim should not fit.

    c. Outside diameter of pipe must be at least 1 in. (2. 54 cm).

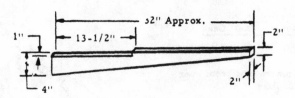

2. Cut stock from wood using a saw or knife.

100

3. Cut a 3/8" deep "V" groove in top of the stock.

4. Turn coupling onto pipe until tight.

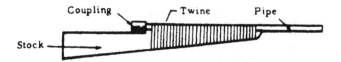

5. Coat pipe and "V" groove of stock with shellac or lacquer and, while still wet, place pipe in "V" groove and wrap pipe and stock together using two heavy layers of twine. Coat twine with shellac or lacquer after each layer.

6. Drill a hole through center of pipe plug large enough for nail to pass through.

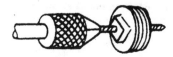

7. File threaded end of plug flat.

8. Push nail through plug and cut off flat 1/32" past the plug.

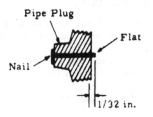

9. Screw plug into coupling.

10. Bend 4" metal strap into "L" shape and drill hole for wood screw. Notch metal strap on the long side 1/2" from bend.

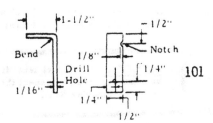

101

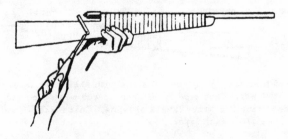

11. Position metal strap on stock so that top will hit the head of the nail. Attach to stock with wood screw.

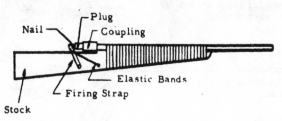

12. Place screw in each side of stock about 4" in front of metal strap. Pass elastic bands through notch in metal strap and attach to screw on each side of the stock.

SAFETY CHECK - TEST FIRE SHOTGUN BEFORE HAND FIRING

1. Locate a barrier such as a stone wall or large tree which you can stand behind in case the weapon explodes when fired.

2. Mount shotgun solidly to a table or other rigid support at least ten feet in front of the barrier.

3. Attach a long cord to the firing strap on the shotgun.

4. Holding the other end of the cord, go behind the barrier.

5. Pull the cord so that the firing strap is held back.

6. Release the cord to fire the shotgun. (If shotgun does not fire, shorten the elastic bands or increase their number.)

IMPORTANT: Fire at least five rounds from behind the barrier and then re-inspect the shotgun before you attempt to shoulder fire it.

HOW TO OPERATE SHOTGUN

1. **To Load**

   a. Take plug out of coupling.

   b. Put shotgun shell into pipe.

   c. Screw plug hand-tight into coupling.

2. **To Fire**

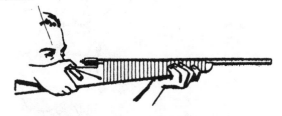

   a. Pull strap back and hold with thumb.

   b. Release strap.

3. **To Unload Gun**

   a. Take plug out of coupling.

   b. Shake out used cartridge.

SHOTSHELL DISPERSION CONTROL

When desired, shotshell can be modified to reduce shot dispersion.

MATERIAL REQUIRED:

Shotshell
Screwdriver or knife
Any of the following filler materials:
    Crushed Rice
    Rice Flour
    Dry Bread Crumbs
    Fine Dry Sawdust

PROCEDURE:

STAR CRIMP

1. Carefully remove crimp from shotshell using a screwdriver or knife.

ROLL CRIMP

NOTE: If cartridge is of roll-crimp type, remove top wad.

2. Pour shot from shell.

3. Replace one layer of shot in the cartridge. Pour in filler material to fill the spaces between the shot.

SHOT

FILLER

WAD

PROPELLANT

104

4. Repeat Step 3 until all shot has been replaced.

5. Replace top wad (if applicable) and re-fold crimp.

6. Roll shell on flat surface to smooth out crimp and restore roundness.

7. Seal end of case with wax.

CANDLE

HOW TO USE:

This round is loaded and fired in the same manner as standard shotshell. The shot spread will be about 2/3 that of a standard round.

CARBINE (7.62 mm Standard Rifle Ammunition)

A rifle can be made from water or gas pipe and fittings. Standard cartridges are used for ammunition.

MATERIAL REQUIRED:

Wood approximately 2 in. x 4 in. x 30 in.

1/4 in. nominal size iron water or gas pipe 20 in. long threaded at one end.

3/8 in. to 1/4 in. reducer

3/8 in. x 1-1/2 in. threaded pipe

3/8 in. pipe coupling

Metal strap approximately 1/2 in. x 1/16 in. x 4 in.

Twine, heavy (100 yards approx.)

3 wood screws and screwdriver

Flat head nail about 1 in. long

Hand drill

Saw or knife

File

Pipe wrench

Shellac or lacquer

Elastic bands

Solid 3/8 in. pipe plug

PROCEDURE:

1. Inspect pipe and fittings carefully.

   a. Be sure that there are no cracks or flaws.

   b. Check inside diameter of pipe. A 7.62 mm projectile should fit into 3/8 in. pipe.

2. Cut stock from wood using saw or knife.

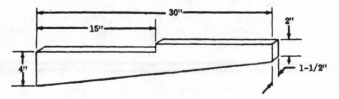

FOR OFFICIAL USE ONLY

3. Cut a 1/4 in. deep "V" groove in top of the stock.

4. Fabricate rifle barrel from pipe.

    a. File or drill inside diameter of threaded end of 20 in. pipe for about 1/4 in. so neck of cartridge case will fit in.

    b. Screw reducer onto threaded pipe using pipe wrench.

    c. Screw short threaded pipe into reducer.

    d. Turn 3/8 pipe coupling onto threaded pipe using pipe wrench. <u>All</u> fittings <u>should</u> be as tight as possible. Do not split fittings.

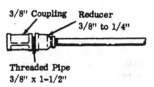

5. Coat pipe and "V" groove of stock with shellac or lacquer. While still wet, place pipe in "V" groove and wrap pipe and stock together using two layers of twine. Coat twine with shellac or lacquer after each layer.

6. Drill a hole through center of pipe plug large enough for nail to pass through.

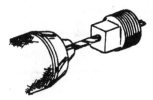

7. File threaded end of plug flat.

107

8. Push nail through plug and cut off rounded 1/32 in. (2 mm) past the plug.

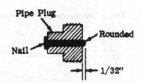

Pipe Plug

Rounded

Nail

1/32"

9. Screw plug into coupling.

10. Bend 4 in. metal strap into "L" shape and drill hole for wood screw. Notch metal strap on the long side 1/2 in. from bend.

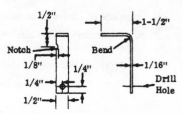

1/2"

1-1/2"

Notch

Bend

1/8"

1/4"

1/16"

1/4"

Drill Hole

1/2"

11. Position metal strap on stock so that top will hit the head of the nail. Attach to stock with wood screw.

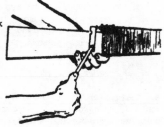

12. Place screw in each side of stock about 4 in. in front of metal strap. Pass elastic bands through notch in metal strap and attach to screw on each side of the stock.

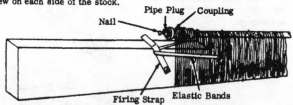

Pipe Plug   Coupling

Nail

Firing Strap   Elastic Bands

SAFETY CHECK - TEST FIRE RIFLE BEFORE HAND FIRING

1. Locate a barrier such as a stone wall or large tree which you can stand behind to test fire weapon.

2. Mount rifle solidly to a table or other rigid support at least ten feet in front of the barrier.

3. Attach a long cord to the firing strap on the rifle.

4. Holding the other end of the cord, go behind the barrier.

5. Pull the cord so that the firing strap is held back.

6. Release the cord to fire the rifle. (If the rifle does not fire, shorten the elastic bands or increase their number.)

> **IMPORTANT:** Fire at least five rounds from behind a barrier and then reinspect the rifle before you attempt to shoulder fire it.

HOW TO OPERATE RIFLE:

1. <u>To Load</u>

   a. Remove plug from coupling.

   b. Put cartridge into pipe.

   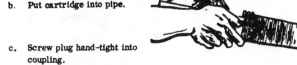

   c. Screw plug hand-tight into coupling.

2. <u>To Fire</u>

   a. Pull strap back and hold with thumb.

   b. Release strap.

3. <u>To Unload Gun</u>

   a. Take plug out of coupling.

   b. Drive out used case using stick or twig.

REUSABLE PRIMER

A method of making a previously fired primer reusable.

MATERIAL REQUIRED:

Used cartridge case
2 long nails having approximately the same diameter as the inside of
    the primer pocket
"Strike-anywhere" matches - 2 or 3 are needed for each primer
Vise
Hammer
Knife or other sharp edged instrument

PROCEDURE:

1.   File one nail to a needle
point so that it is small enough
to fit through hole in primer
pocket.

2.   Place cartridge case and nail between jaws of vise. Force out fired
primer with nail as shown:

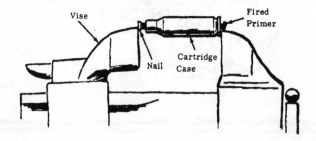

Vise

Fired
Primer

Nail

Cartridge
Case

3.   Remove anvil from primer cup.

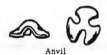

Anvil

4. File down point of second nail until tip is flat.

5. Remove indentations from face of primer cup with hammer and flattened nail.

Nail

Primer Cup

6. Cut off tips of the heads of "strike-anywhere" matches using knife. Carefully crush the match tips on dry surface with wooden match stick until the mixture is the consistency of sugar.

Tip

Head

Wooden Match Stick

CAUTION: Do not crush more than 3 match tips at one time or the mixture may explode.

7. Pour mixture into primer cup. Compress mixture with wooden match stick until primer cup is fully packed.

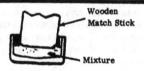

Wooden Match Stick

Mixture

8. Place anvil in primer pocket with legs down.

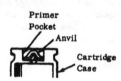

Primer Pocket

Anvil

Cartridge Case

9. Place cup in pocket with mixture facing downward.

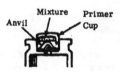

Mixture   Primer Cup

Anvil

10. Place cartridge case and primer cup between vise jaws, and press slowly until primer is seated into bottom of pocket. The primer is now ready to use.

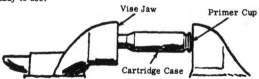

Vise Jaw

Primer Cup

Cartridge Case

111

PIPE PISTOL FOR .45 CALIBER AMMUNITION

A .45 caliber pistol can be made from 3/8 in. nominal diameter steel gas or water pipe and fittings. Lethal range is about 15 yards (13-1/2 meters).

MATERIAL REQUIRED:

Steel pipe, 3/8 in. (1 cm) nominal diameter and 6 in. (15 cm) long
    with threaded ends.
2 threaded couplings to fit pipe
Solid pipe plug to fit pipe coupling
Hard wood, 8-1/2 in. x 6-1/2 in. x 1 in. (21 cm x 16-1/2 cm x 2-1/2 cm)
Tape or string
Flat head nail, approximately 1/16 in. (1-1/2 mm) in diameter
2 wood screws, approximately 1/16 in. (1-1/2 mm) in diameter
Metal strap, 5 in. x 1/4 in. x 1/8 in. (12-1/2 cm x 6 mm x 1 mm)
Bolt, 4 in. (10 cm) long, with nut (optional).
Elastic bands
Drills, one 1/16 in. (1-1/2 mm) in diameter, and one having same
    diameter as bolt (optional).
Rod, 1/4 in. (6mm) in diameter and 8 in. (20 cm) long
Saw or knife

PROCEDURE:

1. Carefully inspect pipe and fittings.

    a. Make sure that there are no cracks or other flaws in the pipe
       and fittings.

    b. Check inside diameter of pipe using a .45 caliber cartridge
       as a gauge. The cartridge case should fit into the pipe snugly
       but without forcing.

    c. Outside diameter of pipe MUST NOT BE less than 1-1/2 times
       the bullet diameter.

2. Follow procedure of Section III, No. 1, steps 4, 5, and 6.

3. Cut stock from wood using saw or knife.

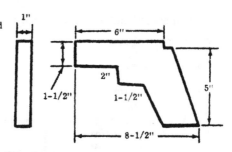

Inches	Centimeters
1-1/2	4 cm
8-1/2	26-1/2
6	20
1-1/2	4
5	12-1/2

4. Cut a 3/8 in. (9-1/2 mm) deep groove in top of stock.

5. Screw couplings onto pipe. Screw plug into one coupling.

6. Securely attach pipe to stock using string or tape.

7. Follow procedures of Section III, No. 1, steps 10 and 11.

8. (Optional) Bend bolt for trigger. Drill hole in stock and place bolt in hole so strap will be anchored by bolt when pulled back. If bolt is not available, use strap as trigger by pulling back and releasing.

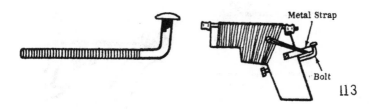

113

9. Follow **SAFETY CHECK**, Section III, No. 1

<u>HOW TO USE</u>:

1. To load:

    a. Remove plug from rear coupling.

    b. Wrap string or elastic band
       around extractor groove so
       case will seat into barrel
       securely.

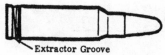

Extractor Groove

    c. Place cartridge in pipe.

    d. Replace plug.

2. <u>To Fire</u>:

    a. Pull metal strap back and
       anchor in trigger.

    b. Pull trigger when ready to fire.

NOTE: If bolt is not used, pull
strap back and release.

3. To remove cartridge case:

    a. Remove plug from rear
       coupling.

    b. Insert rod into front of
       pistol and push cartridge
       case out.

114

MATCH GUN

An improvised weapon using safety match heads as the propellant and a metal object as the projectile. Lethal range is about 40 yards (36 meters).

MATERIAL REQUIRED:

Metal pipe 24 in. (61 cm) long and 3/8 in. (1 cm) in diameter (nominal
    size) or its equivalent, threaded on one end.
End cap to fit pipe
Safety matches - 3 books of 20 matches each.
Wood - 28 in. x 4 in. x 1 in. (70 cm x 10 cm x 2.5 cm)
Toy caps OR safety fuse OR "Strike-anywhere matches" (2)
Electrical tape or string
Metal strap, about 4 in. x 1/4 in. x 3/16 in. (10 cm x 6 mm x 4.5 mm)
2 rags, about 1 in x 12 in. and 1 in. x 3 in. (2-1/2 cm x 30 cm and
    2-1/2 cm x 8 cm)
Wood screws
Elastic bands
Metal object (steel rod, bolt with head cut off, etc.), approximately
    7/16 in. (11 mm) in diameter, and 7/16 in. (11 mm) long if iron
    or steel, 1-1/4 in. (31 mm) long if aluminum, 5/16 in. (8mm) long
    if lead.
Metal disk 1 in. (2-1/2 cm) in diameter and 1/16 in. (1-1/2 mm) thick
Bolt, 3/32 in. (2-1/2 mm) or smaller in diameter and nut to fit
Saw or knife

PROCEDURE:

1. Carefully inspect pipe and fittings. Be sure that there are no cracks or other flaws.

2. Drill small hole in center
of end cap. If safety fuse is used,
be sure it will pass through this
hole.

	Metric	English
	5 cm	2 in.
3. Cut stock from wood using saw or knife.	10 cm	4 in.
	36 cm	14 in.
	71 cm	28 in.

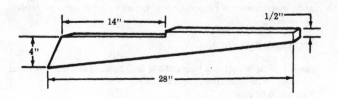

4. Cut 3/8 in. (9-1/2 mm) deep "V" groove in top of stock.

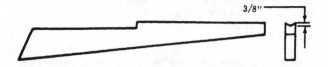

5. Screw end cap onto pipe until finger tight.

6. Attach pipe to stock with string or tape.

7. Bend metal strap into "L" shape and drill holes for wood screw. Notch metal strap on long side 1/2 in. (1 cm) from bend.

8. Position metal strap on stock so that the top will hit the center of hole drilled in end cap.

9. Attach metal disk to strap with
nut and bolt. This will deflect blast
from hole in end cap when gun is
fired. Be sure that head of bolt is
centered on hole in end cap.

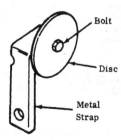

Bolt

Disc

Metal
Strap

10. Attach strap to stock with wood screws.

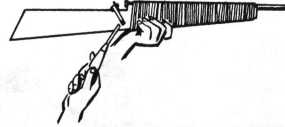

11. Place screw on each side of stock about 4 in. (10 cm) in front of
metal strap. Pass elastic bands through notch in metal strap and attach
to screw on each side of stock.

HOW TO USE:

A. When Toy Caps Are Available:

1. Cut off match heads from 3
books of matches with knife.
Pour match heads into pipe.

2. Fold one end of 1 in. x 12 in. rag 3 times so that it becomes a one inch square of 3 thicknesses. Place rag into pipe to cover match heads, folded end first. Tamp firmly WITH CAUTION.

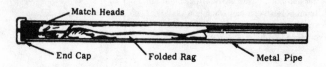

Match Heads

End Cap   Folded Rag   Metal Pipe

3. Place metal object into pipe. Place 1 in. x 3 in. rag into pipe to cover projectile. Tamp firmly WITH CAUTION.

4. Place 2 toy caps over small hole in end cap. Be sure metal strap will hit caps when it is released.

NOTE: It may be necessary to tape toy caps to end cap.

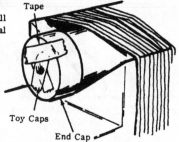

Tape

Toy Caps

End Cap

5. When ready to fire, pull metal strap back and release.

B. When "Strike-Anywhere" Matches Are Available:

1. Follow steps 1 through 3 in A.

Tip
Head
Wooden
Match Stick

2. Carefully cut off tips of heads of 2 "strike-anywhere" matches with knife.

3. Place one tip in hole in end cap. Push in with wooden end of match stick.

118

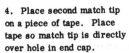

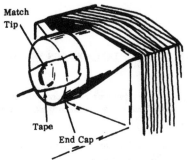

4. Place second match tip on a piece of tape. Place tape so match tip is directly over hole in end cap.

Match Tip

Tape

End Cap

5. When ready to fire, pull metal strap back and release.

C. When Safety Fuse Is Available: (Recommended for Booby Traps)

1. Remove end cap from pipe. Knot one end of safety fuse. Thread safety fuse through hole in end cap so that knot is on <u>inside</u> of end cap.

2. Follow steps 1 through 3 in A.

3. Tie several matches to safety fuse near outside of end cap.

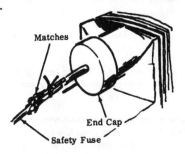

Matches

End Cap

Safety Fuse

NOTE: Bare end of safety fuse should be inside match head cluster.

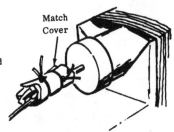

Match Cover

4. Wrap match covers around matches and tie. Striker should be in contact with match bands.

5. Replace end cap on pipe.

6. When ready to fire, pull match cover off with strong, firm, quick motion.

119

SAFETY CHECK - TEST FIRE GUN BEFORE HAND FIRING

1. Locate a barrier such as a stone wall or large tree which you can stand behind in case the weapon explodes when fired.

2. Mount gun solidly to a table or other rigid support at least ten feet in front of the barrier.

3. Attach a long cord to the firing strap on the gun.

4. Holding the other end of the cord, go behind the barrier.

5. Pull the cord so that the firing strap is held back.

6. Release the cord to fire the gun. (If gun does not fire, shorten the elastic bands or increase their number.)

IMPORTANT: Fire at least five rounds from behind the barrier and then re-inspect the gun before you attempt to shoulder fire it.

RIFLE CARTRIDGE

NOTE: See Section III, No. 5 for reusable primer.

A method of making a previously fired rifle cartridge reusable.

MATERIAL REQUIRED:

Empty rifle cartridge, be sure that it still fits inside gun.
Threaded bolt that fits into neck of cartridge at least 1-1/4 in. (3 cm)
   long.
Safety or "strike-anywhere" matches (about 58 matches are needed
   for 7.62 mm cartridge)
Rag wad (about 3/4 in. (1-1/2 cm) square for 7.62 mm cartridge)
Knife
Saw

NOTE: Number of matches and size of rag wad depend on particular
cartridge used.

PROCEDURE:

1. Remove coating on heads of
matches by scraping match sticks
with sharp edge.

CAUTION: If wooden "strike-anywhere" matches are used, cut off
tips <u>first</u>. Discard tips or use for
Reusable Primer, Section III, No. 5.

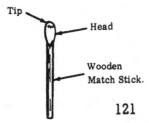

Tip

Head

Wooden
Match Stick.

121

2. Fill previously primed cartridge case with match head coatings up to its neck. Pack evenly and tightly with match stick.

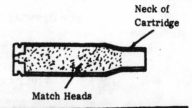

Neck of Cartridge

Match Heads

CAUTION: Remove head of match stick before packing. In all packing operations, stand off to the side and pack gently. Do not hammer.

3. Place rag wad in neck of case. Pack with match stick from which head was removed.

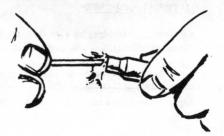

4. Saw off head end of bolt so remainder is approximately the length of the standard bullet.

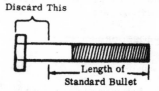

Discard This

Length of Standard Bullet

5. Place bolt in cartridge case so that it sticks out about the same length as the original bullet.

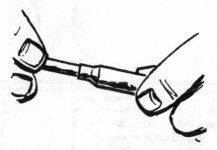

NOTE: If bolt does not fit snugly, force paper or match sticks between bolt and case, or wrap tape around bolt before inserting in case.

122

FOR OFFICIAL USE ONLY

PIPE PISTOL FOR .38 CALIBER AMMUNITION

A .38 caliber pistol can be made from 1/4 in. nominal diameter steel gas or water pipe and fittings. Lethal range is approximately 33 yards (30 meters).

MATERIAL REQUIRED:

Steel pipe, 1/4 in. (6 mm) nominal
    diameter and 6 in. (15 cm)
    long with threaded ends (nipple)
Solid pipe plug, 1/4 in. (6 mm)
    nominal diameter
2 steel pipe couplings, 1/4 in. (6 mm)
    nominal diameter
Metal strap, approximately 1/8 in.
    x 1/4 in. x 5 in. (3 mm x 6 mm
    x 125 mm or 12-1/2 cm)
Elastic bands
Flat head nail - 6D or 8D, approxi-
    mately 1/16 in. diameter
    (1-1/2 mm)
2 wood screws, #8
Hard wood, 8 in. x 5 in. x 1 in.
    (20 cm x 12-1/2 cm x 2-1/2 cm)
Drill
Wood or metal rod, 1/4 in. (6 mm)
    diameter and 8 in. (20 cm) long
Saw or knife

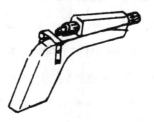

PROCEDURE:

1. Carefully inspect pipe and fittings.

    a. Make sure that there are NO cracks or other flaws in the pipe or fittings

    b. Check inside diameter of pipe using a .38 caliber cartridge as a gauge. The bullet should fit closely into the pipe without forcing, but the cartridge case SHOULD NOT fit into the pipe.

    c. Outside diameter of pipe MUST NOT BE less than 1-1/2 times the bullet diameter.

2. Drill a 35/64 in. (14 mm) di-. ameter hole 3/4 in. (2 cm) into one coupling to remove the thread. Drilled section should fit tightly over smooth section of pipe.

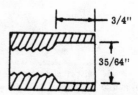

3. Drill a 25/64 in. (1 cm) diam- eter hole 1-1/8 in. (2.86 cm) into pipe. Use cartridge as a guage; when a cartridge is inserted into the pipe, the shoulder of the case should butt against the end of the pipe. Thread coupling tightly onto pipe, drilled end first.

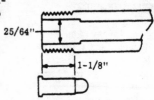

4. Follow procedures of Section III, No. 1, steps 4 through 11.

5. Follow SAFETY CHECK, Section III, No. 1.

HOW TO OPERATE PISTOL:

Follow procedures of HOW TO OPERATE PISTOL, Section III, No. 1, steps 1, 2, and 3.

PIPE PISTOL FOR .22 CALIBER AMMUNITION
LONG OR SHORT CARTRIDGE

A .22 Caliber pistol can be made from 1/8 in. nominal diameter extra heavy, steel gas or water pipe and fittings. Lethal range is approximately 33 yards (30 meters).

MATERIAL REQUIRED:

Steel pipe, extra heavy, 1/8 in.
    (3 mm) nominal diameter and
    6 in. (15 cm) long with
    threaded ends (nipple)
Solid pipe plug, 1/8 in. (3 mm)
    nominal diameter
2 steel pipe couplings, 1/8 in. (3 mm)
    nominal diameter
Metal strap, approximately 1/8 in.
    x 1/4 in. x 5 in. (3 mm x 6 mm
    x 125 mm or 12-1/2 cm)
Elastic bands
Flat head nail - 6D or 8D (approxi-
    mately 1/16 in. (1-1/2 mm)
    diameter
2 wood screws, #8
Hard wood, 8 in. x 5 in. x 1 in.
    (20 cm x 12-1/2 cm x 2-1/2 cm)
Drill
Wood or metal rod, 1/8 in. (3 mm)
    diameter and 8 in. (20 cm) long
Saw or knife

PROCEDURE:

1. Carefully inspect pipe and fittings.

    a. Make sure that there are NO cracks or other flaws in the pipe
        or fittings.

    b. Check inside diameter of pipe using a .22 caliber cartridge,
        long or short, as a gauge. The bullet should fit closely into
        the pipe without forcing, but the cartridge case SHOULD NOT
        fit into the pipe.

    c. Outside diameter of pipe MUST NOT BE less than 1-1/2 times
        the bullet diameter.

2. Drill a 15/64 in. (1/2 cm) diameter hole 9/16 in. (1-1/2 cm)
deep in pipe for long cartridge. (If
a short cartridge is used, drill hole
3/8 in. (1 cm) deep). When a cartridge is inserted into the pipe, the
shoulder of the case should butt
against the end of the pipe.

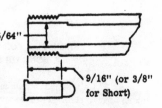

15/64"

9/16" (or 3/8"
for Short)

3. Screw the coupling onto the pipe. Cut coupling length to allow pipe
plug to thread in pipe flush against the cartridge case.

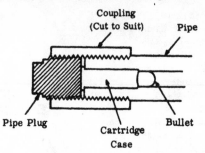

Coupling
(Cut to Suit)

Pipe

Pipe Plug

Cartridge
Case

Bullet

4. Drill a hole off center of the
pipe plug just large enough for the
nail to fit through.

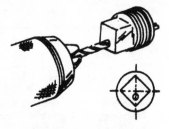

NOTE: Drilled hole MUST BE OFF
CENTER in plug.

5. Push nail through pipe plug until
head of nail is flush with square end.
Cut nail off at other end 1/16 in.
(1-1/2 mm) away from plug. Round
off end with file.

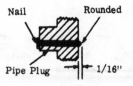

Nail

Rounded

Pipe Plug

1/16"

6. Follow procedures of Section III, No. 1, steps 6 through 11.

7. Follow SAFETY CHECK, Section III, No. 1.

HOW TO OPERATE:

126 Follow procedures of HOW TO OPERATE PISTOL, Section III, No. 1,
steps 1, 2, and 3.

LOW SIGNATURE SYSTEM

Low signature systems (silencers) for improvised small arms weapons (Section III) can be made from steel gas or water pipe and fittings.

MATERIAL REQUIRED:

Grenade container
Steel pipe nipple, 6 in. (15 cm) long –
    See Table I for diameter
2 steel pipe couplings - See Table II
    for dimensions
Cotton cloth - See Table II for
    dimensions
Drill
Absorbent cotton

PROCEDURE:

1. Drill hole in grenade container at both ends to fit outside diameter of pipe nipple. (See Table I.)

2. Drill four (4) rows of holes in pipe nipple. Use Table I for diameter and location of holes.

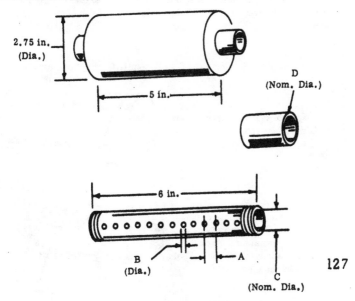

2.75 in. (Dia.)

5 in.

D (Nom. Dia.)

6 in.

B (Dia.)

A

C (Nom. Dia.)

127

Table I. Low Signature System Dimensions

	A	B	C	(Coupling) D	Holes per Row	(4-Rows) Total
.45 Cal.	3/8	1/4	3/8	3/8	12	48
.38 Cal.	3/8	1/4	1/4	1/4	12	48
9 mm	3/8	1/4	1/4	1/4	12	48
7.62 mm	3/8	1/4	1/4	1/4	12	48
.22 Cal.	1/4	5/32	1/8*	1/8	14	50

*Extra Heavy Pipe
All dimensions in inches

3. Thread one of the pipe couplings on the drilled pipe nipple.

Cut

4. Cut coupling length to allow barrel of weapon to thread fully into low signature system. Barrel should butt against end of the drilled pipe nipple.

Top Half   Bottom Half

5. Separate the top half of the grenade container from the bottom half.

Grenade Container

6. Insert the pipe nipple in the drilled hole at the base of the bottom half of container. Pack the absorbent cotton inside the container and around the pipe nipple.

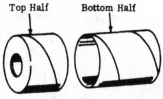

Drilled Pipe Nipple

Coupling

Absorbent Cotton

128

7. Pack the absorbent cotton in top half of grenade container leaving hole in center. Assemble container to the bottom half.

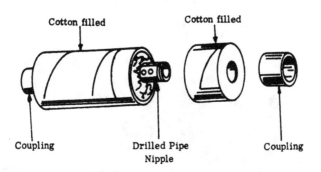

Cotton filled   Cotton filled

Coupling   Drilled Pipe Nipple   Coupling

8. Thread the other coupling onto the pipe nipple.

NOTE: A longer container and pipe nipple, with same "A" and "B" dimensions as those given, will further reduce the signature of the system.

HOW TO USE:

1. Thread the low signature system on the selected weapon securely.

2. Place the proper cotton wad size into the muzzle end of the system.

Table II. Cotton Wadding - Sizes

Weapon	Cotton Wad Size
.45 Cal.	1-1/2 x 6 inches
.38 Cal.	1 x 4 inches
9 mm	1 x 4 inches
7.62 mm	1 x 4 inches
.22 Cal.	Not needed

3. Load Weapon

4. Weapon is now ready for use.

RECOILLESS LAUNCHER

A dual directional scrap fragment launcher which can be placed to cover the path of advancing troops.

MATERIAL REQUIRED:

Iron water pipe approximately 4 ft. (1 meter) long and 2 to 4 in. (5 to 10 cm) in diameter

Black powder (commercial) or salvaged artillery propellant about 1/2 lb. (200 gms)

Safety or improvised fuse (Section VI, No. 7) or improvised electrical igniter (Section VI, No. 2)

Stones and/or metal scrap chunks approximately 1/2 in. (1 cm) in diameter - about 1 lb. (400 gms) total

4 rags for wadding, each about 20 in. by 20 in. (50 cm by 50 cm)

Wire

Paper or rag

NOTE: Be sure that the water pipe has no cracks or flaws.

PROCEDURE:

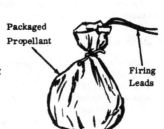

1. Place propellant and igniter
in paper or rag and tie with string
so contents cannot fall out.

2. Insert packaged propellant and igniter in center of pipe. Pull firing
leads out one end of pipe.

3. Stuff a rag wad into each end of pipe and lightly tamp using a flat
end stick.

4. Insert stones and/or scrap metal into each end of pipe. Be sure
the same weight of material is used in each side.

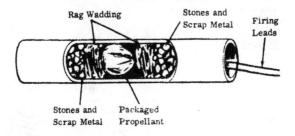

5. Insert a rag wad into each end of the pipe and pack tightly as before.

HOW TO USE:

1. Place scrap mine in a tree or pointed in the path of the enemy.
Attach igniter lead to the firing circuit. The recoilless launcher is
now ready to fire.

2. If safety or improvised fuse is used instead of the detonator, place
the fuse into the packaged propellant through a hole drilled in the center
of the pipe. Light free end of fuse when ready to fire. Allow for normal
delay time.

CAUTION: Scrap will be ejected from both ends of the launcher.

131

SHOTGUN GRENADE LAUNCHER

This device can be used to launch a hand grenade to a distance of 160 yards (150 meters) or more, using a standard 12 gauge shotgun.

MATERIAL REQUIRED:

Grenade (Improvised pipe hand grenade, Section II, No. 1, may be used)
12 gauge shotgun
12 gauge shotgun cartridges
Two washers, (brass, steel, iron, etc.), having outside diameter of 5/8 in. (1-1/2 cm)
Rubber disk 3/4 in. (2 cm) in diameter and 1/4 in. (6 mm) thick (leather, neoprene, etc. can be used)
A 30 in. (75 cm) long piece of hard wood (maple, oak, etc.) approximately 5/8 in. (1-1/2 cm) in diameter. Be sure that wood will slide into barrel easily.
Tin can (grenade and its safety lever must fit into can)
Two wooden blocks about 2 in. (5 cm) square and 1-1/2 in. (4 cm) thick
One wood screw about 1 in. (2-1/2 cm) long
Two nails about 2 in. (5 cm) long
12 gauge wads, tissue paper, or cotton
Adhesive tape, string, or wire
Drill

PROCEDURE:

1.   Punch hole in center of rubber disk large enough for screw to pass through.

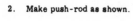

2.   Make push-rod as shown.

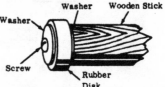

NOTE: Gun barrel is slightly less than 3/4 inch in diameter. If rubber disk does not fit in barrel, file or trim it very slightly. It should fit tightly.

3.   Drill a hole through the center of one wooden block of such size that the push-rod will fit tightly. Whittle a depression around the hole on one side approximately 1/8 in. (3 mm) and large enough for the grenade to rest in.

132

4. Place the base of the grenade
in the depression in the wooden
block. Securely fasten grenade to
block by wrapping tape (or wire)
around entire grenade and block.

NOTE: Be sure that the tape (or
wire) does not cover hole in block
or interfere with the operation of
the grenade safety lever.

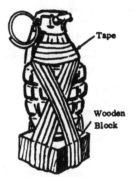

Tape

Wooden
Block

5. Drill hole through the center of the second wooden block, so that it
will just slide over the outside of the gun barrel.

6. Drill a hole in the center of the bottom of the tin can the same size
as the hole in the block.

7. Attach can to block as shown.

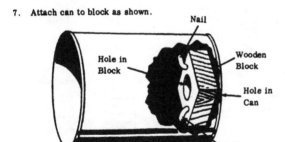

Nail

Hole in
Block

Wooden
Block

Hole in
Can

Nail

8. Slide the can and block onto the barrel until muzzle passes can
open end. Wrap a small piece of tape around the barrel an inch or
two from the end. Tightly wrapped string may be used instead of tape.
Force the can and wooden block forward against the tape so that they
are securely held in place. Wrap tape around the barrel behind the can.

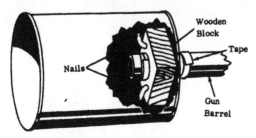

Wooden
Block

Tape

Nails

Gun
Barrel

CAUTION: Be sure that the can is securely fastened to the gun barrel. If the can should become loose and slip down the barrel after the launcher is assembled. the grenade will explode after the regular delay time.

9. Remove crimp from a 12 gauge shotgun cartridge with pen knife. Open cartridge. Pour shot from shell Remove wads and plastic liner if present.

10. Empty the propellant onto a piece of paper. Using a knife. divide the propellant in half. Replace half of the propellant into the cartridge case.

11. Replace the 12 gauge cardboard wads into cartridge case.

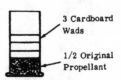

3 Cardboard Wads

1/2 Original Propellant

NOTE: If wads are not available, stuff tissue paper or cotton into the cartridge case. Pack tightly.

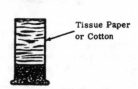

Tissue Paper or Cotton

HOW TO USE:

Method I - When ordinary grenade is used:

1. Load cartridge in gun.

2. Push end of push-rod without the rubber disk into hole in wooden block fastened to grenade.

3. Slowly push rod into barrel until it rests against the cartridge case and grenade is in can. If the grenade is not in the can, remove rod and cut to proper size. Push rod back into barrel.

Grenade

4. With can holding safety lever of grenade in place, carefully remove safety pin.

CAUTION: Be sure that the sides of the can restrain the grenade safety lever. If the safety lever should be released for any reason, grenade will explode after regular grenade delay time.

5. To fire grenade launcher, rest gun in ground at angle determined by range desired. A 45 degree angle should give about 150 meters (160 yds.).

Method II - When improvised pipe grenade is used:

An improvised pipe grenade (Section II, No. 1) may be launched in a similar manner. No tin can is needed.

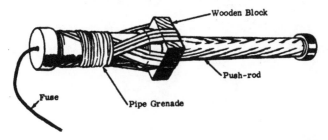

Wooden Block

Push-rod

Fuse

Pipe Grenade

1. Fasten the grenade to the block as shown above with the fuse hole at the end opposite the block.

2. Push end of push-rod into hole in wooden block fastened to grenade.

3. Push rod into barrel until it rests against cartridge case.

4. Load cartridge in gun.

5. Follow step 5 of Method I.

6. Using a fuse with <u>at least</u> a 10 second delay, light the fuse before firing.

7. Fire when the fuse burns to 1/2 its original length.

FOR OFFICIAL USE ONLY

GRENADE LAUNCHER (57 MM CARDBOARD CONTAINER)

An improvised method of launching a standard grenade 150 yds. (135 meters) or an improvised grenade 90 yds. (81 meters) using a discarded cardboard ammunition container.

MATERIAL REQUIRED:

Heavy cardboard container with inside diameter of 2-1/2 to 3 in.
   (5-1/2 to 8 cm) and at least 12 in. (30 cm) long (ammunition
   container is suitable)
Black powder - 8 grams (124 grains) or less
Safety or improvised fuse (Section VI, No. 7)
Grenade (Improvised hand grenade, Section II, No. 1 may be used)
Rag, approximately 30 in. x 24 in. (75 cm x 60 cm)
Paper

> CAUTION: 8 grams of black powder yield the maximum ranges. Do not use more than this amount. See Improvised Scale, Section VII, No. 8, for measuring.

PROCEDURE: METHOD I - If Standard Grenade is Used.

Top of
Container

1. Discard top of container.
Make small hole in bottom.

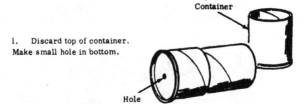

Hole

2. Place black powder in paper.
Tie end with string so contents
cannot fall out. Place package in
container.

3. Insert rag wadding into container. Pack tightly with CAUTION.

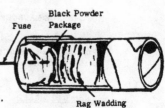

4. Measure off a length of fuse
that will give the desired delay.
Thread this through hole in bottom of container so that it pene-
trates into the black powder package.

NOTE: If improvised fuse is used, be sure fuse fits loosely through
hole in bottom of container.

5. Hold grenade safety lever and
carefully withdraw safety pin from
grenade. Insert grenade into container, lever end first.

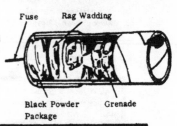

> CAUTION: If grenade safety lever should be released for any reason,
> grenade will explode after normal delay time.

6.   Bury container about 6 in. (15 cm) in the ground at 30° angle, bring-
ing fuse up alongside container. Pack ground tightly around container.

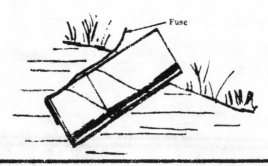

> CAUTION: The tightly packed dirt helps to hold the tube together dur-
> ing the firing. Do not fire unless at least the bottom half of the container
> is buried in solidly packed dirt.

METHOD II - If Improvised Pipe Hand Grenade is Used.

1.   Follow step 1 of above procedure.

2. Measure off a piece of fuse at least as long as the cardboard container. Tape one end of this to the fuse from the blasting cap in the improvised grenade. Be sure ends of fuse are in contact with each other.

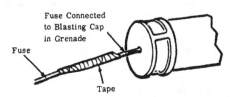

Fuse Connected to Blasting Cap in Grenade

Fuse

Tape

3. Place free end of fuse and black powder on piece of paper. Tie ends with string so contents will not fall out.

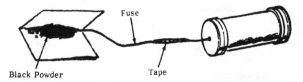

Fuse

Black Powder

Tape

4. Place package in tube. Insert rag wadding. Pack so it fits snugly. Place pipe hand grenade into tube. Be sure it fits snugly.

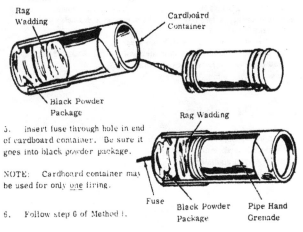

Rag Wadding

Cardboard Container

Black Powder Package

Rag Wadding

5. Insert fuse through hole in end of cardboard container. Be sure it goes into black powder package.

NOTE: Cardboard container may be used for only one firing.

Fuse   Black Powder Package   Pipe Hand Grenade

6. Follow step 6 of Method 1.

HOW TO USE:

Light fuse when ready to fire.

FOR OFFICIAL USE ONLY

FIRE BOTTLE LAUNCHER

A device using 2 items (shotgun and chemical fire bottle) that can be used to start or place a fire 80 yards (72 meters) from launcher.

MATERIAL REQUIRED:

Standard 12 gauge or improvised shotgun (Section III, No. 2)
Improvised fire bottle (Section V, No. 1)
Tin can, about 4 in. (10 cm) in diameter and 5-1/2 in. (14 cm) high
Wood, about 3 in. x 3 in. x 2 in. (7-1/2 cm x 7-1/2 cm x 5 cm)
Nail, at least 3 in. (7-1/2 cm) long
Nuts and bolts or nails, at least 2-1/2 in. (6-1/2 cm) long
Rag
Paper
Drill

If Standard Shotgun is Used:

Hard wood stick, about the same length as shotgun barrel and about
    5/8 in. (1-1/2 cm) in diameter. Stick need not be round.
2 washers (brass, steel, iron, etc.) having outside diameter of 5/8
    in. (1-1/2 cm)
One wood screw about 1 in. (2-1/2 cm) long
Rubber disk, 3/4 in. (2 cm) in diameter and 1/4 in. (6 mm) thick,
    leather, cardboard, etc. can be used.
12 gauge shotgun ammunition

If Improvised Shotgun is Used:

Fuse, safety or improvised fast burning (Section VI, No. 7)
Hard wood stick, about the same length as shotgun barrel and 3/4
    in. (2 cm) in diameter
Black powder - 9 grams (135 grains). See Section VII, No. 8.

PROCEDURE:

METHOD I - If Improvised Shotgun is Used:

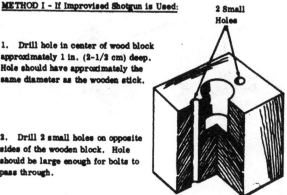

2 Small
Holes

1. Drill hole in center of wood block approximately 1 in. (2-1/2 cm) deep. Hole should have approximately the same diameter as the wooden stick.

2. Drill 2 small holes on opposite sides of the wooden block. Hole should be large enough for bolts to pass through.

3. Fasten can to block with nuts and bolts.

NOTE: Can may also be securely fastened to block by hammering several nails through can and block. Do not drill holes, and be careful not to split wood.

4. Place wooden stick into hole in wooden block. Drill small hole (same diameter as that of 3 in. nail) through wooden block and through wooden stick. Insert nail in hole.

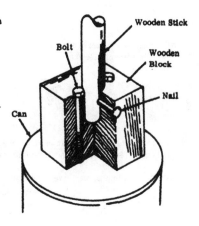

Wooden Stick

Bolt

Wooden
Block

Nail

Can

5. Crumple paper and place in bottom of can. Place another piece of paper around fire bottle and insert in can. Use enough paper so that bottle will fit snugly.

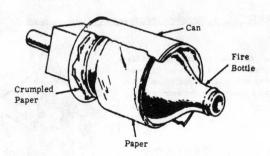

6. Place safety fuse and black powder on paper. Tie each end with string.

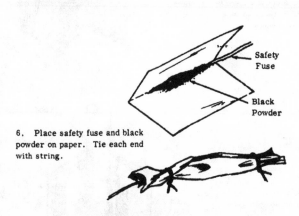

7. Thread fuse through hole in plug. Place powder package in rear of shotgun. Screw plug finger tight into coupling.

NOTE: Hole in plug may have to be enlarged for fuse.

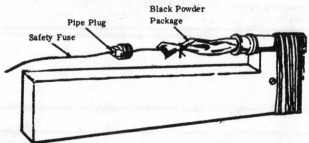

142

8. Insert rag into front of shotgun. Pack rag against powder package with stick. USE CAUTION.

METHOD II - If Standard Shotgun is Used:

1. Follow Steps 1 and 2, Shotgun Grenade Launcher, Section IV, No. 2.

2. Follow procedure of Method I, Steps 1 - 5.

3. Follow Steps 9, 10, 11, Shotgun Grenade Launcher, Section IV, No. 2, using 1/3 of total propellant instead of 1/2.

4. Load cartridge in gun.

HOW TO USE:

1. Insert stick and holder containing chemical fire bottle.

CAUTION: Do not tilt muzzle downward.

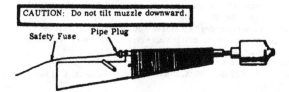

Safety Fuse   Pipe Plug

2. Hold gun against ground at 45° angle and light fuse.

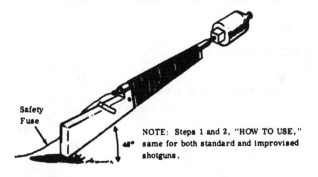

Safety
Fuse

NOTE: Steps 1 and 2, "HOW TO USE," same for both standard and improvised shotguns.

45°

CAUTION: Severe burns may result if bottle shatters when fired. If possible, obtain a bottle identical to that being used as the fire bottle. Fill about 2/3 full of water and fire as above. If bottle shatters when fired instead of being launched intact, use a different type of bottle.

143

Section IV
No. 5

GRENADE LAUNCHERS

A variety of grenade launchers can be fabricated from metal pipes and fittings. Ranges up to 600 meters (660 yards) can be obtained depending on length of tube, charge, number of grenades, and angle of firing.

MATERIAL REQUIRED:

Metal pipe, threaded on one end and approximately 2-1/2 in. (6-1/4 cm) in diameter and 14 in. to 4 ft. (35 cm to 119 cm) long depending on range desired and number of grenades used.

End cap to fit pipe

Black powder, 15 to 50 gm, approximately 1-1/4 to 4-1/4 tablespoons (Section I, No. 3)

Safety fuse, fast burning improvised fuse (Section VI, No. 7) or improvised electric bulb initiator (Section VI, No. 1 Automobile light bulb is needed)

Grenade(s) - 1 to 6

Rag(s) - about 30 in. x 30 in. (75 cm x 75 cm) and 20 in. x 20 in. (55 cm x 55 cm)

Drill

String

NOTE: Examine pipe carefully to be sure there are no cracks or other flaws.

PROCEDURE:

METHOD I - If Fuse is Used:

1. Drill small hole through center of end cap.

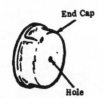

End Cap

Hole

2. Make small knot near one end of fuse. Place black powder and knotted end of fuse in paper and tie with string.

Fuse

Knot

Black Powder

String

144

3. Thread fuse through hole in end cap and place package in end cap. Screw end cap onto pipe, being careful that black powder package is not caught between the threads.

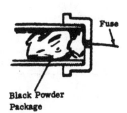

Fuse

Black Powder Package

4. Roll rag wad so that it is about 6 in. (15 cm) long and has approximately the same diameter as the pipe. Push rolled rag into open-end of pipe until it rests against black powder package.

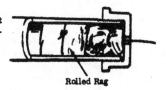

Rolled Rag

5. Hold grenade safety lever in place and carefully withdraw safety pin.

CAUTION: If grenade safety lever is released for any reason, grenade will explode after regular time. (4 - 5 sec.)

6. Holding safety lever in place, carefully push grenade into pipe, lever end first, until it rests against rag wad.

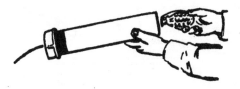

145

7.   The following table lists various types of grenade launchers and their performance characteristics.

DESIRED RANGE	NO. OF GRENADES LAUNCHED	BLACK POWDER CHARGE	PIPE LENGTH	FIRING ANGLE
250 m	1	15 gm	14"	30°
500 m	1	50 gm	48"	10°
600 m [a]	1	50 gm	48"	30°
200 m	6 [b]	25 gm	48"	30°

(a)   For this range, an additional delay is required. See Section VI, No. 11 and 12.

(b)   For multiple grenade launcher, load as shown.

NOTE: Since performance of different black powder varies, fire several test rounds to determine the exact amount of powder necessary to achieve the desired range.

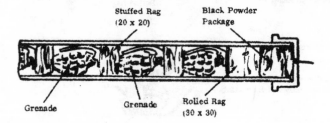

HOW TO USE:

1.   Bury at least 1/2 of the launcher pipe in the ground at desired angle. Open end should face the expected path of the enemy. Muzzle may be covered with cardboard and a thin layer of dirt and/or leaves as camouflage. Be sure cardboard prevents dirt from entering pipe.

NOTE: The 14 in. launcher may be hand held against the ground instead of being buried.

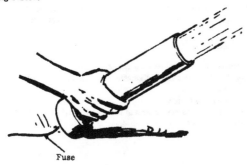

Fuse

2. Light fuse when ready to fire.

METHOD II - If Electrical Igniter is Used:

NOTE: Be sure that bulb is in good operating condition.

1. Prepare electric bulb initiator as described in Section VI, No. 1.

2. Place electric initiator and black powder charge in paper. Tie ends of paper with string.

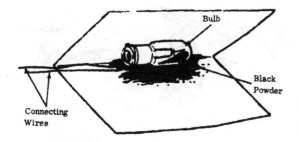

Bulb

Black Powder

Connecting Wires

3. Follow above Procedure, Steps 3 to end.

HOW TO USE:

1. Follow above How to Use, Step 1.

2. Connect leads to firing circuit. Close circuit when ready to fire.

147

60 MM MORTAR PROJECTILE LAUNCHER

A device to launch 60 mm mortar rounds using a metal pipe 2-1/2 in. (6 cm) in diameter and 4 ft. (120 cm) long as the launching tube.

MATERIAL REQUIRED:

Mortar, projectile (60 mm) and charge increments
Metal pipe 2-1/2 in. (6 cm) in diameter and 4 ft. (120 cm) long, threaded
    on one end
Threaded end cap to fit pipe
Bolt, 1/8 in. (3 mm) in diameter and at least 1 in. (2-1/2 cm) long
Two (2) nuts to fit bolt
File
Drill

PROCEDURE:

1.   Drill hole 1/8 in. (3 mm) in diameter through center of end cap.

2.   Round off end of bolt with file.

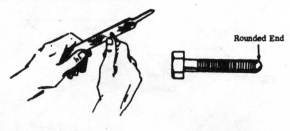

3.   Place bolt through hole in end cap. Secure in place with nuts as illustrated.

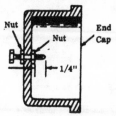

4. Screw end cap onto pipe tightly. Tube is now ready for use.

HOW TO USE:

1. Bury launching tube in ground at desired angle so that bottom of tube is at least 2 ft. (60 cm) underground. Adjust the number of increments in rear finned end of mortar projectile. See following table for launching angle and number of increments used.

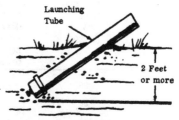

Launching Tube

2 Feet or more

2. When ready to fire, withdraw safety wire from mortar projectile. Drop projectile into launching tube, FINNED END FIRST.

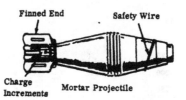

Finned End   Safety Wire

Charge Increments   Mortar Projectile

CAUTION: Be sure bore riding pin is in place in fuse when mortar projectile is dropped into tube. A live mortar round could explode in the tube if the fit is loose enough to permit the bore riding pin to come out partway.

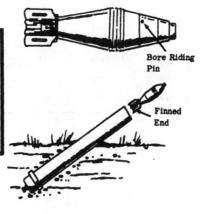

Bore Riding Pin

Finned End

CAUTION: The round will fire as soon as the projectile is dropped into tube. Keep all parts of body behind the open end of the tube.

149

DESIRED RANGE (YARDS)	MAXIMUM HEIGHT MORTAR WILL REACH (YARDS)	REQUIRED ANGLE OF ELEVATION OF TUBE (MEASURED FROM HORIZONTAL DEGREES)	CHARGE - NUMBER OF INCREMENTS
150	25	40	0
300	50	40	1
700	150	40	2
1000	225	40	3
1500	300	40	4
125	75	60	0
300	125	60	1
550	250	60	2
1000	375	60	3
1440	600	60	4
75	100	80	0
150	200	80	1
300	350	80	2
400	600	80	3
550	750	80	4

CHEMICAL FIRE BOTTLE

This incendiary bottle is self-igniting on target impact.

MATERIALS REQUIRED

	How Used	Common Source
Sulphuric Acid	Storage Batteries Material Proces- sing	Motor Vehicles Industrial Plants
Gasoline	Motor Fuel	Gas Station or Motor Vehicles
Potassium Chlorate	Medicine	Drug Store
Sugar	Sweetening Foods	Food Store

Glass bottle with stopper (roughly 1 quart size).
Small Bottle or jar with lid.
Rag or absorbent paper (paper towels, newspaper).
String or rubber bands.

PROCEDURE

1. **Sulphuric Acid Must be Concentrated.** If battery acid or other dilute acid is used, concentrate it by boiling until dense white fumes are given off. Container used should be of enamel-ware or oven glass.

CAUTION

Sulphuric acid will burn skin and destroy clothing. If any is spilled, wash it away with a large quantity of water. Fumes are also dangerous and should not be inhaled.

151

2. Remove the acid from heat and allow to cool to room temperature.

3. Pour gasoline into the large (1 quart) bottle until it is approximately 2/3 full.

4. Add concentrated sulphuric acid to gasoline slowly until the bottle is filled to within 1" to 2" from top. Place the stopper on the bottle.

5. Wash the outside of the bottle thoroughly with clear water.

CAUTION

If this is not done, the fire bottle may be dangerous to handle during use.

---

6. Wrap a clean cloth or several sheets of absorbent paper around the outside of the bottle. Tie with string or fasten with rubber bands.

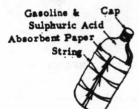

Gasoline & Cap
Sulphuric Acid
Absorbent Paper
String

7. Dissolve 1/2 cup (100 gm) of potassium chlorate and 1/2 cup (100 gm) of sugar in one cup (250 cc) of boiling water.

8. Allow the solution to cool, pour into the small bottle and cap tightly. The cooled solution should be approx. 2/3 crystals and 1/3 liquid. If there is more liquid than this, pour off excess before using.

CAUTION

Store this bottle separately from the other bottle.

HOW TO USE

1. Shake the small bottle to mix contents and pour onto the cloth or paper around the large bottle.

152 Bottle can be used wet or after solution has dried. However, when dry, the sugar - Potassium chlorate mixture is very sensitive to spark or flame and should be handled accordingly.

2. Throw or launch the bottle. When the bottle breaks against a hard surface (target) the fuel will ignite.

IGNITER FROM BOOK MATCHES

This is a hot igniter made from paper book matches for use with molotov cocktail and other incendiaries.

Material Required

Paper book matches.
Adhesive or friction tape.

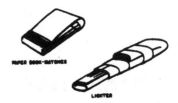

Procedure

1. Remove the staple(s) from match book and separate matches from cover.

2. Fold and tape one row of matches.

3. Shape the cover into a tube with striking surface on the inside and tape. Make sure the folded cover will fit tightly around the taped match heads. Leave cover open at opposite end for insertion of the matches.

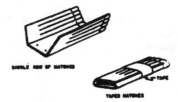

4. Push the taped matches into the tube until the bottom ends are exposed about 3/4 in. (2 cm).

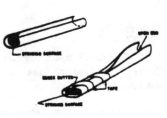

153

5. Flatten and fold the open end of the tube so that it laps over about 1 in. (2-1/2 cm); tape in place.

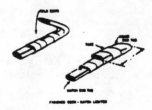

Use With Molotov Cocktail

Tape the "match end tab" of the igniter to the neck of the molotov cocktail.

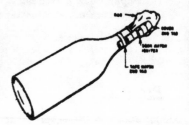

Grasp the "cover end tab" and pull sharply or quickly to ignite.

General Use

The book match igniter can be used by itself to ignite flammable liquids, fuse cords and similar items requiring hot ignition.

CAUTION

Store matches and completed igniters in mois-tureproof containers such as rubber or plastic bags until ready for use. Damp or wet paper book matches will not ignite.

154

MECHANICALLY INITIATED FIRE BOTTLE

The mechanically initiated Fire Bottle is an incendiary device which ignites when thrown against a hard surface.

MATERIALS REQUIRED

Glass jar or short neck bottle with
a leakproof lid or stopper.
"Tin" can or similar container just
large enough to fit over the lid
of the jar.
Coil spring (compression) approxi-
mately 1/2 the diameter of the
can and 1 1/2 times as long.
Gasoline
Four (4) "blue tip" matches
Flat stick or piece of metal
(roughly 1/2" x 1/16" x 4")
Wire or heavy twine
Adhesive tape

PROCEDURE

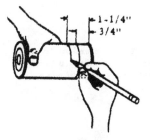

1. Draw or scratch two lines
around the can - one 3/4" (19
mm) and the other 1 1/4" (30
mm) from the open end.

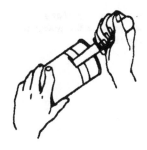

2. Cut 2 slots on opposite sides
of the tin can at the line farthest
from the open end. Make slots
large enough for the flat stick or
piece of metal to pass through.

155

3. Punch 2 small holes just
below the rim of the open end of
the can.

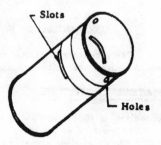

4. Tape blue tip matches together
in pairs. The distance between
the match heads should equal the
inside diameter of the can. Two
pairs are sufficient.

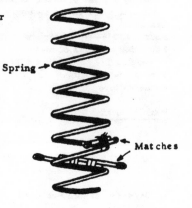

5. Attach paired matches to
second and third coils of the
spring, using thin wire.

6. Insert the end of the
spring opposite the matches
into the tin can.

7. Compress the spring until the end with the matches passes the slot in the can. Pass the flat stick or piece of metal through slots in can to hold spring in place. This acts as a safety device.

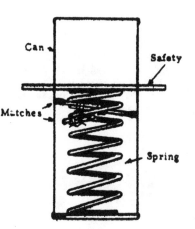

8. Punch many closely spaced small holes between the lines marked on the can to form a striking surface for the matches. Be careful not to seriously deform can.

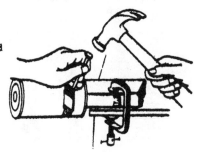

9. Fill the jar with gasoline and cap tightly.

10. Turn can over and place over the jar so that the safety stick rests on the lid of the jar.

11. Pass wire or twine around the bottom of the jar. Thread ends through holes in can and bind tightly to jar.

12. Tape wire or cord to jar near the bottom.

HOW TO USE

1. Carefully withdraw flat safety stick.

2. Throw jar at hard surface.

CAUTION:

DO NOT REMOVE SAFETY STICK UNTIL READY TO THROW FIRE BOTTLE.

The safety stick, when in place, prevents ignition of the fire bottle if it should accidentally be broken.

GELLED FLAME FUELS

Gelled or paste type fuels are often preferable to raw gasoline for use in incendiary devices such as fire bottles. This type fuel adheres more readily to the target and produces greater heat concentration.

Several methods are shown for gelling gasoline using commonly available materials. The methods are divided into the following categories based on the major ingredient:

4.1 Lye Systems

4.2 Lye-Alcohol Systems

4.3 Soap-Alcohol Systems

4.4 Egg White Systems

4.5 Latex Systems

4.6 Wax Systems

4.7 Animal Blood Systems

GELLED FLAME FUELS

LYE SYSTEMS

Lye (also known as caustic soda or Sodium Hydroxide) can be used in combination with powdered rosin or castor oil to gel gasoline for use as a flame fuel which will adhere to target surfaces.

NOTE: This fuel is not suitable for use in the chemical (Sulphuric Acid) type of fire bottle (Section V, No.1). The acid will react with the lye and break down the gel.

MATERIALS REQUIRED:

Parts by Volume	Ingredient	How Used	Common Source
60	Gasoline	Motor fuel	Gas station or motor vehicle
2 (flake) or 1 (powder)	Lye	Drain cleaner, making of soap	Food store Drug store
15	Rosin	Manufacturing Paint & Varnish	Naval stores Industry
	or		
	Castor Oil	Medicine	Food and Drug Stores

PROCEDURE:

CAUTION: Make sure that there are no open flames in the area when mixing the flame fuel. NO SMOKING!

1. Pour gasoline into jar, bottle or other container. (DO NOT USE AN ALUMINUM CONTAINER.)

2. If rosin is in cake form, crush into small pieces.

3. Add rosin or castor oil to the gasoline and stir for about five (5) minutes to mix thoroughly.

4. In a second container (NOT ALUMINUM) add lye to an equal volume of water slowly with stirring.

CAUTION: Lye solution can burn skin and destroy clothing. If any is spilled, wash away immediately with large quantities of water.

5. Add lye solution to the gasoline mix and stir until mixture thickens (about one minute).

160 NOTE: The sample will eventually thicken to a very firm paste. This can be thinned, if desired, by stirring in additional gasoline.

GELLED FLAME FUELS

LYE-ALCOHOL SYSTEMS

Lye (also known as caustic soda or Sodium Hydroxide) can be used in combination with alcohol and any of several fats to gel gasoline for use as a flame fuel.

NOTE: This fuel is not suitable for use in the chemical (Sulphuric Acid) type of fire bottle (Section V, No. 1). The acid will react with the lye and break down the gel.

MATERIALS REQUIRED:

Parts by Volume	Ingredient	How Used	Common Source
60	Gasoline	Motor fuel	Gas station or motor vehicles
2 (flake) or 1 (powder)	Lye	Drain cleaner Making of soap	Food store Drug store
3	Ethyl Alcohol	Whiskey Medicine	Liquor store Drug store

NOTE: Methyl (wood) alcohol or isopropyl (rubbing) alcohol can be substituted for ethyl alcohol, but their use produces softer gels.

14	Tallow	Food Making of soap	Fat rendered by cooking the meat or suet of animals.

NOTE: The following can be substituted for the tallow:

    (a) Wool grease (Lanolin) (very good). -- Fat extracted from sheep wool.
    (b) Castor oil (good).
    (c) Any vegetable oil (corn, cottonseed, peanut, linseed, etc.)
    (d) Any fish oil
    (e) Butter or oleomargarine

It is necessary when using substitutes (c) to (e) to double the given amount of fat and of lye for satisfactory bodying.

PROCEDURE:

CAUTION: Make sure that there are no open flames in the area when mixing flame fuels. NO SMOKING!

1. Pour gasoline into bottle, jar or other container. (DO NOT USE AN ALUMINUM CONTAINER).

2. Add Tallow (or substitute) to the gasoline and stir for about 1/2 minute to dissolve fat.

161

3. Add alcohol to the gasoline mixture.

4. In a separate container (NOT ALUMINUM) slowly add lye to an equal amount of water. Mixture should be stirred constantly while adding lye.

CAUTION: Lye solution can burn skin and destroy clothing. If any is spilled, wash away immediately with large quantities of water.

5. Add lye solution to the gasoline mixture and stir occasionally until thickened (about 1/2 hour).

NOTE: The mixture will eventually (1 to 2 days) thicken to a very firm paste. This can be thinned, if desired, by stirring in additional gasoline.

GELLED FLAME FUELS

SOAP-ALCOHOL SYSTEM

Common household soap can be used in combination with alcohol to gel gasoline for use as a flame fuel which will adhere to target surfaces.

MATERIAL REQUIRED:

Parts by Volume	Ingredient	How Used	Common Source
36	Gasoline	Motor fuel	Gas station, Motor vehicles
1	Ethyl Alcohol	Whiskey Medicine	Liquor store Drug store

NOTE: Methyl (wood) or isopropyl (rubbing) alcohols can be substituted for the whiskey.

20 (powdered) or 28 (flake)	Laundry soap	Washing clothes	Stores

NOTE: Unless the word "soap" actually appears somewhere on the container or wrapper, a washing compound is probably a detergent. These Can Not Be Used.

PROCEDURE:

CAUTION: Make sure that there are no open flames in the area when mixing flame fuels. NO SMOKING!

1. If bar soap is used, carve into thin flakes using a knife.

2. Pour alcohol and gasoline into a jar, bottle or other container and mix thoroughly.

3. Add soap powder or flakes to gasoline-alcohol mix and stir occasionally until thickened (about 15 minutes).

163

GELLED FLAME FUELS

EGG SYSTEMS

The white of any bird egg can be used to gel gasoline for use as a flame fuel which will adhere to target surfaces.

MATERIALS REQUIRED:

Parts by Volume	Ingredient	How Used	Common Source
85	Gasoline	Motor fuel Stove fuel Solvent	Gas station Motor vehicles
14	Egg Whites	Food Industrial processes	Food store Farms

Any One Of The Following:

1	Table Salt	Food Industrial processes	Sea water Natural brine Food store
3	Ground Coffee	Food	Coffee plant Food store
3	Dried Tea Leaves	Food	Tea plant Food store
3	Cocoa	Food	Cacao tree Food store
2	Sugar	Sweetening foods Industrial processes	Sugar cane Food store
1	Saltpeter (Niter) (Potassium Nitrate)	Pyrotechnics Explosives Matches Medicine	Natural Deposits Drug store
1	Epsom salts	Medicine Mineral water Industrial processes	Natural deposits Kieserite Drug store Food store
2	Washing soda (Sal soda)	Washing cleaner Medicine Photography	Food store Drug store Photo supply store

Parts by Volume	Ingredient	How Used	Common Source
1 1/2	Baking Soda	Baking Manufacture of: Beverages, Mineral waters and Medicines	Food store Drug store
1 1/2	Aspirin	Medicine	Drug store Food store

PROCEDURE:

CAUTION: Make sure that there are no open flames in the area when mixing flame fuels. NO SMOKING!

1. Separate egg white from yolk. This can be done by breaking the egg into a dish and carefully removing the yolk with a spoon.

NOTE: DO NOT GET THE YELLOW EGG YOLK MIXED INTO THE EGG WHITE. If egg yolk gets into the egg white, discard the egg.

2. Pour egg white into a jar, bottle, or other container and add gasoline.

3. Add the salt (or other additive) to the mixture and stir occasionally until gel forms (about 5 to 10 minutes).

NOTE: A thicker gelled flame fuel can be obtained by putting the capped jar in hot (65°C) water for about 1/2 hour and then letting them cool to room temperature. (DO NOT HEAT THE GELLED FUEL CONTAINING COFFEE).

GELLED FLAME FUELS

LATEX SYSTEMS

Any milky white plant fluid is a potential source of latex which can be used to gel gasoline

MATERIALS REQUIRED:

Ingredient	How Used	Common Source
Gasoline	Motor fuel Solvent	Gas station Motor vehicle
Latex, commerical or natural	Paints Adhesives	Natural from tree or plant Rubber cement

One of the Following Acids:

Acetic Acid (Vinegar)	Salad dressing Developing film	Food stores Fermented apple cider Photographic supply
Sulfuric Acid (Oil of Vitriol)	Storage batteries Material processing	Motor vehicles Industrial plants
Hydrochloric Acid (Muriatic Acid)	Petroleum wells Pickling and metal cleaning Industrial processes	Hardware store Industrial plants

NOTE: If acids are not available, use acid salt (alum, sulfates and chlorides other than sodium or potassium). The formic acid from crushed red ants can also be used.

PROCEDURE:

> CAUTION: Make sure that there are no open flames in the area when mixing flame fuels. NO SMOKING!

1. With Commercial Rubber Latex:

a. Place 7 parts by volume of latex and 92 parts by volume of gasoline in bottle. Cap bottle and shake to mix well.

b. Add 1 part by volume vinegar (or other acid) and shake until gel forms.

> CAUTION: Concentrated acids will burn skin and destroy clothing. If any is spilled, wash away immediately with large quantities of water.

2. With Natural Latex:

a. Natural latex should form lumps as it comes from the plant. If lumps do not form, add a small amount of acid to the latex.

b. Strain off the latex lumps and allow to dry in air.

c. Place 20 parts by volume of latex in bottle and add 80 parts by volume of gasoline. Cover bottle and allow to stand until a swollen gel mass is obtained (2 to 3 days).

Section V
No. 4.6

WAX SYSTEMS

Any of several common waxes can be used to gel gasoline for use as a flame fuel which will adhere to target surfaces.

MATERIALS REQUIRED:

Parts by Volume	Ingredient	How Used	Common Source
80	Gasoline	Motor fuel Solvent	Gas station Motor vehicles

Any one of the following:

20	Ozocerite Mineral wax Fossil wax Ceresin wax	Leather polish Sealing wax Candles Crayons Waxed paper Textile sizing	Natural deposits General stores Department store
	Beeswax	Furniture and floor waxes Artificial fruit and flowers Lithographing Wax paper Textile finish Candles	Honeycomb of bee General store Department store
	Bayberry wax Myrtle wax	Candles Soaps Leather polish Medicine	Natural form Myrica berries General store Department store Drug store

PROCEDURE:

1. Obtaining wax from Natural Sources: Plants and berries are potential sources of natural waxes. Place the plants and/or berries in boiling water. The natural waxes will melt. Let the water cool. The natural waxes will form a solid layer on the water surface. Skim off the solid wax and let it dry. With natural waxes which have suspended matter when melted, screen the wax through a cloth.

2. Melt the wax and pour into jar or bottle which has been placed in a hot water bath.

3. Add gasoline to the bottle.

4. When wax has completely dissolved in the gasoline, allow the water bath to cool slowly to room temperature.

NOTE: If a gel does not form, add additional wax (up to 40% by volume) and repeat the above steps. If no gel forms with 40% wax, make a Lye solution by dissolving a small amount of Lye (Sodium Hydroxide) in an equal amount of water. Add this solution (1/2% by volume) to the gasoline wax mix and shake bottle until a gel forms

168

GELLED FLAME FUELS

ANIMAL BLOOD SYSTEMS

Animal blood can be used to gel gasoline for use as a flame fuel which will adhere to target surfaces.

MATERIAL REQUIRED:

Parts by Volume	Ingredient	How Used	Common Source
68	Gasoline	Motor fuel Solvent	Gas station Motor vehicles
30	Animal blood Serum	Food Medicine	Slaughter House Natural habitat
Any one of the following:			
2	Salt	Food Industrial processes	Sea Water Natural brine Food store
	Ground Coffee	Food Caffeine source Beverage	Coffee plant Food store
	Dried Tea Leaves	Food Beverage	Tea plant Food store
	Sugar	Sweetening foods Industrial processes	Sugar cane Food store
	Lime	Mortar Plaster Medicine Ceramics Steel making Industrial processes	From calcium carbonate Hardware store Drug store Garden supply store
	Baking soda	Baking Beverages Medicine Industrial processes	Food store Drug store
	Epsom salts	Medicine Mineral water Industrial processes	Drug store Natural deposits Food store

169

PROCEDURE:

1. Preparation of animal blood serum:

   a. Slit animal's throat by jugular vein. Hang up-side down to drain.

   b. Place coagulated (lumpy) blood in a cloth or on a screen and catch the red fluid (serum) which drains through.

   c. Store in cool place if possible.

CAUTION: Do not get aged animal blood or the serum into an open cut. This can cause infections.

2. Pour blood serum into jar, bottle, or other container and add gasoline.

3. Add the salt (or other additive) to the mixture and stir until a gel forms.

ACID DELAY INCENDIARY

This device will ignite automatically after a given time delay.

MATERIAL REQUIRED:

Small jar with cap
Cardboard
Adhesive tape
Potassium Chlorate
Sugar
Sulphuric Acid (Battery Acid)
Rubber sheeting (automotive inner tube)

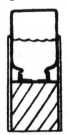

PROCEDURE:

1. Sulphuric acid must be concentrated. If battery acid or other dilute acid is used, concentrate it by boiling. Container used should be of enamelware or oven glass. When dense white fumes begin to appear, immediately remove the acid from heat and allow to cool to room temperature.

> CAUTION: Sulphuric acid will burn skin and destroy clothing. If any is spilled, wash it away with a large quantity of water. Fumes are also dangerous and should not be inhaled.

2. Dissolve one part by volume of Potassium Chlorate and one part by volume of sugar in two parts by volume of boiling water.

3. Allow the solution to cool. When crystals settle, pour off and discard the liquid.

JAR

4. Form a tube from cardboard just large enough to fit around the outside of the jar and 2 to 3 times the height of the jar. Tape one end of the tube closed.

CARDBOARD

POTASSIUM
CHLORATE-
SUGAR

5. Pour wet Potassium Chlorate-sugar crystals into the tube until it is about 2/3 full. Stand the tube aside to dry.

CARBOARD
TUBE

6. Drill a hole through the cap of the jar about 1/2 inch (1 1/4 cm) in diameter.

171

JAR
LID

7. Cut a disc from rubber sheet so that it just fits snugly inside the lid of the jar.

RUBBER SHEET

8. Partly fill jar with water, cover with rubber disc and cap tightly with the drilled lid. Invert bottle and allow to stand for a few minutes to make sure that there are no leaks. THIS IS EXTREMELY IMPORTANT.

9. Pour water from jar and fill about 1/3 full with concentrated sulphuric acid. Replace the rubber disc and cap tightly.

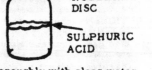

CAP

RUBBER DISC

SULPHURIC ACID

IMPORTANT: Wash outside of jar thoroughly with clear water. If this is not done, the jar may be dangerous to handle during use.

HOW TO USE:

1. Place the tube containing the Sugar Chlorate crystals on an incendiary or flammable material taped end down.

2. Turn the jar of sulphuric acid cap end down and slide it into the open end of the tube.

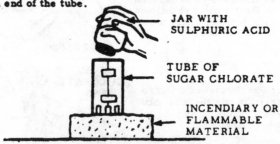

JAR WITH SULPHURIC ACID

TUBE OF SUGAR CHLORATE

INCENDIARY OR FLAMMABLE MATERIAL

After a time delay, the acid will eat through the rubber disc and ignite the sugar chlorate mix. The delay time depends upon the thickness and type of rubber used for the disc. Before using this device, tests should be conducted to determine the delay time that can be expected.

NOTE: A piece of standard automobile inner tube (about 1/32" thick) will provide a delay time of approximately 45 minutes.

172

FOR OFFICIAL USE ONLY

IMPROVISED WHITE FLARE

An improvised white flare can be made from potassium nitrate, aluminum powder and shellac. It has a time duration of approximately 2 minutes.

MATERIALS REQUIRED:	SOURCE:
Potassium nitrate	Field grade (Section I, No. 2)
	Drug Store
Aluminum powder (bronzing)	Hardware or paint store
Shellac	Hardware or paint store
Quart jar with lid	
Fuse, 15 in. long	
Wooden rod, 1/4 in. diameter	
Tin can, 2-1/2 in. diameter x 5 in. long	
Flat window screen	
Wooden block	

NOTE: All of the above dimensions are approximate.

PROCEDURE:

1. Place the potassium nitrate crystals on the screen. Rub the material back and forth against the screen mesh with the wooden block until the nitrate is granulated into a powder.

2. Measure 21 tablespoons of the powdered nitrate into a quart jar. Add 21 tablespoons of the aluminum powder to the nitrate.

Potassium Nitrate

173

3. Place lid on the jar and shake ingredients vigorously until well mixed.

4. Add 12 tablespoons of shellac to the mixture and stir with the wooden rod. Store mixture until ready for Step 7.

5. Knot one end of the fuse.

6. Wrap the knotted end of the fuse once around the inside bottom of the can with the knot at the center. Then, run the rest of the fuse out the center top of the can.

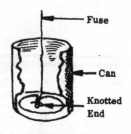

7. Pour the mixture in the can and around the fuse.

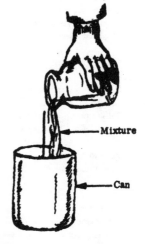

Mixture

Can

8. Store flare mixture away from heat and flame until ready for use, but no longer than 3 weeks.

IMPROVISED IRON OXIDE

Iron Oxide can be made from steel wool. It is used in the preparation of Improvised Yellow Flare (Section V, No. 8), Improvised White Smoke Munition (Section V, No. 9) and Improvised Black Smoke Munition (Section V, No. 10).

MATERIAL REQUIRED:	SOURCE:
Steel wool (without soap), approx. 16 large pads	Hardware or general store
Smoke pipe, approximately 4 feet long x 12 inches in diameter, 1/16 inches thick	Hardware store
Vacuum cleaner	Hardware store
Electrical source (110 v., A.C.)	Modern commercial and domestic buildings

Window screen
Newspaper
2 containers
Wooden blocks, if necessary
Flame source (matches, lighter, etc.)

PROCEDURE:

1. Separate a handful of steel wool into a fluffy ball approximately 12 inches in diameter and place into one end of the smoke pipe.

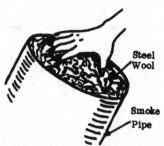

Steel Wool

Smoke Pipe

2. Place the pipe on a level, nonflammable surface. Steady the pipe, using wooden blocks if necessary.

3. Ignite the steel wool with the flame source and, with the vacuum cleaner, force a stream of air through the flame.

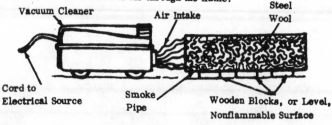

Vacuum Cleaner   Air Intake   Steel Wool

Cord to Electrical Source   Smoke Pipe   Wooden Blocks, or Level, Nonflammable Surface

NOTE: The forced air provided by the vacuum cleaner aids in the burning of the steel wool. If the steel wool does not completely burn, more separation of the wool is needed.

4. When the steel wool has almost completely burned, add another handful of the fluffed steel wool (Step No. 1).

5. Continue adding to the flame a single handful of fluffed wool at a time until a sufficient amount of iron oxide granules have accumulated in the stove pipe.

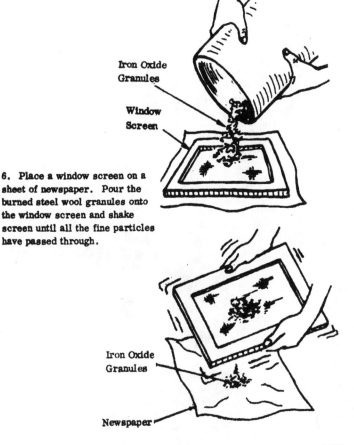

Iron Oxide Granules

Window Screen

6. Place a window screen on a sheet of newspaper. Pour the burned steel wool granules onto the window screen and shake screen until all the fine particles have passed through.

Iron Oxide Granules

Newspaper

177

7. Discard those particles on the newspaper which are fibrous and unburned.

8. Save the particles which were too large to pass through the screen in one of the containers for future burning.

9. Store particles of iron oxide (left on newspaper) in another container until ready for use.

IMPROVISED YELLOW FLARE

A yellow flare can be made from shellac, sulfur, aluminum powder, iron oxide and baking soda. It can be used either for signaling or lighting up a dark area.

MATERIALS REQUIRED:	SOURCES:
Shellac	Hardware or paint store
Sulfur	Drug or agricultural supply store
Aluminum powder (bronzing)	Hardware or paint store
Black iron oxide	Section V, No. 7
Sodium bicarbonate (baking soda)	Food store
Improvised white flare mix	Section V, No. 6
Window Screen	
Wooden rod or stick	
Tablespoon	
Quart jar with lid	
Newspaper	
Wooden block	
Fuse, 15 inches long	
Tin can, 2-1/2 inches diameter x 5 inches long	
Aluminum foil	
Flame source (matches, lighter, etc.)	

PROCEDURE:

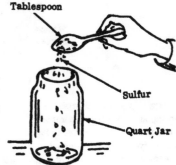

Tablespoon

Sulfur

Quart Jar

1. Measure 6 firm level tablespoons of sulfur into a quart jar.

2. Add 7 firm level tablespoons of sodium bicarbonate to the sulfur.

3. Add 2 heaping tablespoons of black iron oxide.

179

Quart Jar Lid

4. Place the lid on the quart jar
and shake ingredients 10 times.

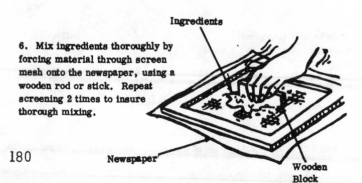

Ingredients

5. Place the mixed ingredients on
the window screen.

Window
Screen

Ingredients

6. Mix ingredients thoroughly by
forcing material through screen
mesh onto the newspaper, using a
wooden rod or stick. Repeat
screening 2 times to insure
thorough mixing.

180   Newspaper

Wooden
Block

7. Pour mixed ingredients back into the jar.

Mixed
Ingredients

8. Add 20 heaping tablespoons of aluminum powder to the ingredients.

Aluminum
Powder

9. Add while stirring the least amount of shellac needed to moisten mixture.

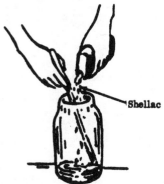

Shellac

181

10. Force moistened mix through screen mesh onto the newspaper as in Step 6. Store mixture until ready for Step 14.

11. Measure one heaping teaspoon of white flare mix onto a 4 inch square piece of aluminum foil.

12. Knot one end of the fuse and place the knot onto the mix.

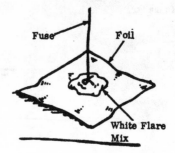

13. Fold the corners of the foil tightly around the fuse.

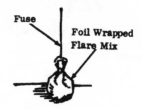

14. Now place the yellow flare mix into the can.

182

Fuse

White Flare
Mix in Foil

Yellow Flare
Mix

15. Place the fused white flare mix in the foil below the surface of the yellow flare mix in the can.

16. Light the fuse with the flame source when ready.

IMPROVISED WHITE SMOKE MUNITION

A white smoke munition can be made from sulfur, potassium nitrate, black powder, aluminum powder, iron oxide and carbon tetrachloride. It can be used either for signaling or screening.

MATERIAL REQUIRED:	SOURCE:
Sulfur	Drug or agricultural supply store
Potassium nitrate (Saltpeter)	Drug store or Section I, No. 2
Improvised black powder	Section I, No. 3
Aluminum powder (bronzing)	Hardware or paint store
Black iron oxide	Section V, No. 7
Carbon tetrachloride	Hardware or paint store
Improvised white flare mix	Section V, No. 6
Tablespoon	
Wooden rod or stick	
Newspaper	
Quart jar with lid	
Window screen	
Fuse, 15 inches long	
Tin can, 2-1/2 inches diameter	
x 5 inches long	
Flame source (matches, lighter,	
etc.)	

PROCEDURE:

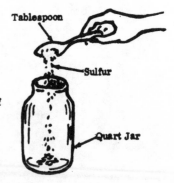

Tablespoon

Sulfur

Quart Jar

1. Measure 3 level tablespoons of powdered dry sulfur into the quart jar.

2. Add 4 level tablespoons of powdered dry potassium nitrate to the sulfur.

NOTE: It may be necessary to crush the potassium nitrate crystals and sulfur to obtain an accurate measure in tablespoon.

3. Add 2 heaping tablespoons of black iron oxide.

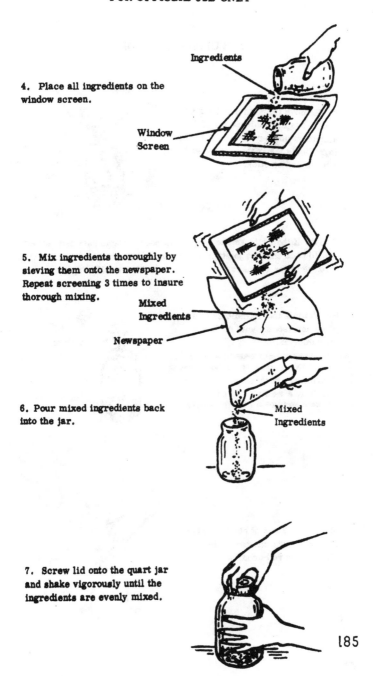

4. Place all ingredients on the window screen.

Ingredients

Window Screen

5. Mix ingredients thoroughly by sieving them onto the newspaper. Repeat screening 3 times to insure thorough mixing.

Mixed Ingredients

Newspaper

6. Pour mixed ingredients back into the jar.

Mixed Ingredients

7. Screw lid onto the quart jar and shake vigorously until the ingredients are evenly mixed.

185

8. Remove lid from quart jar and add 15 heaping tablespoons of aluminum powder (bronzing) to the ingredients. Mix thoroughly with wooden rod or stick.

Wooden rod or stick

Mixed Ingredients plus Aluminum Powder

NOTE: If the white smoke mixture is not for immediate use, screw the lid back onto the jar tightly and store until ready for use. If mixture is for immediate use, continue with the following steps.

9. Wet mix the ingredients to a paste consistency with carbon tetrachloride.

CAUTION: Fumes of Carbon Tetrachloride are hazardous. Perform Step 10 in a well ventilated area.

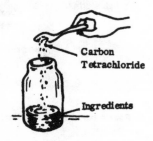

Carbon Tetrachloride

Ingredients

10. Add 1/2 cup of black powder to the ingredients and carefully mix with wooden rod or stick.

Black Powder

HOW TO USE:

1. Measure one heaping teaspoon of white flare mix onto a 4 inch square piece of aluminum foil.

2. Knot one end of the fuse and place the knot into the mix.

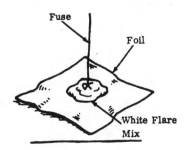

Fuse

Foil

White Flare Mix

3. Fold the corners of the foil tightly around the fuse.

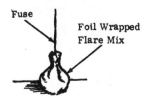

Fuse

Foil Wrapped Flare Mix

4. Now place the white smoke mix into the can.

White Smoke Mix

5. Place the fused white flare mix in the foil below the surface of the white smoke mix in the can.

6. Light the fuse with the flame source when ready.

IMPROVISED BLACK SMOKE MUNITION

A black smoke munition can be made from sulfur, aluminum powder, iron oxide, moth crystals and carbon tetrachloride. It can be used either for signaling or screening.

MATERIAL REQUIRED: SOURCES:

Sulfur   Drug store
Aluminum powder (bronzing)   Paint or hardware store
Improvised black iron oxide   Section V, No. 7
Moth crystals (paradichloroben-   Hardware store
  zene)
Carbon tetrachloride   Paint or hardware store
Improvised white flare mix   Section V, No. 6
Table salt   Food store
Teaspoon
Tablespoon
Quart jar or container
Wooden rod or stick
Wooden block
Window screen
Newspaper
Fuse, 15 in. long
Tin can, 2-1/2 in. diameter x 5 in.
  long
Aluminum foil
Flame source (matches, lighter, etc.)

PROCEDURE:

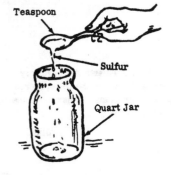

Teaspoon

Sulfur

Quart Jar

1. Measure 3 level teaspoons of sulfur into a quart jar.

2. Add 1 heaping tablespoon of improvised iron oxide to the sulfur.

3. Add 2 level teaspoons of table salt.

189

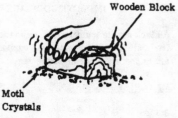

Wooden Block

4. Crush 5 heaping tablespoons of moth crystal into a fine powder using a wooden block.

Moth
Crystals

5. Add 4 heaping tablespoons of powdered moth crystals to the other ingredients in jar.

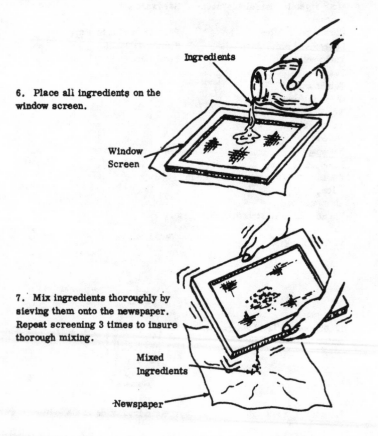

Ingredients

6. Place all ingredients on the window screen.

Window
Screen

7. Mix ingredients thoroughly by sieving them onto the newspaper. Repeat screening 3 times to insure thorough mixing.

Mixed
Ingredients

Newspaper

190

8. Pour mixed ingredients back into the jar.

Mixed Ingredients

9. Add 12 heaping tablespoons of aluminum powder to the ingredients and mix by stirring with wooden rod or stick.

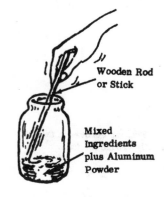

Wooden Rod or Stick

Mixed Ingredients plus Aluminum Powder

10. Just before use as a black smoke, wet mix the above ingredients to a paste consistency with carbon tetrachloride.

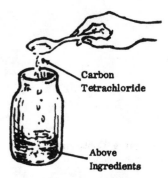

Carbon Tetrachloride

Above Ingredients

CAUTION: Fumes of Carbon Tetrachloride are hazardous. Perform Step 10 in a well ventilated area.

191

HOW TO USE:

1. Measure one heaping teaspoon of white flare mix onto a 4 inch square piece of aluminum foil.

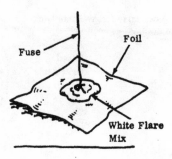

2. Knot one end of the fuse and place the knot into the mix.

Fuse

Foil

White Flare Mix

3. Fold the corners of the foil tightly around the fuse.

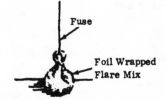

Fuse

Foil Wrapped Flare Mix

4. Now place the black smoke mix into the can.

Black Smoke Mix (Paste)

192

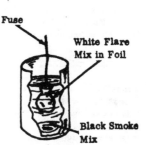

Fuse

White Flare
Mix in Foil

Black Smoke
Mix

5. Place the fused white flare mix in the foil below the surface of the black smoke mix in the can.

6. Light the fuse with the flame source when ready.

ELECTRIC BULB INITIATOR

Mortars, mines and similar weapons often make use of electric initiators. An electric initiator can be made using a flashlight or automobile electric light bulb.

MATERIAL REQUIRED

Electric light bulb and
   mating socket
Cardboard or heavy paper
Black Powder
Adhesive tape

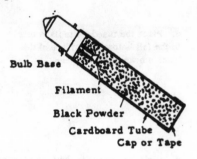

Bulb Base

Filament

Black Powder

Cardboard Tube

Cap or Tape

PROCEDURE

Method I

1. Break the glass of the electric light bulb. Take care not to damage the filament. The initiator will NOT work if the filament is broken. Remove all glass above the base of the bulb.

2. Form a tube 3 to 4 inches long from cardboard or heavy paper to fit around the base of the bulb. Join the tube with adhesive tape.

3. Fit the tube to the bulb base and tape in place.

Make sure that the tube does not cover that portion of the bulb base that fits into the socket.

194

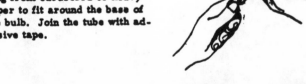

Filament

Cardboard Tube

Tape

Bulb Base

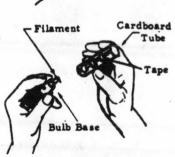

4. If no socket is available for connecting the initiator to the firing circuit, solder the connecting wires to the bulb base.

CAUTION: Do NOT use a hot soldering iron on the completed igniter since it may ignite the Black Powder.

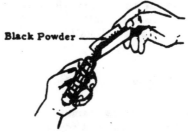

Black Powder

5. Fill the tube with Black Powder and tape the open end of the tube closed.

Method II

If the glass bulb (electric light) is large enough to hold the Black Powder, it can be used as the container.

PROCEDURE

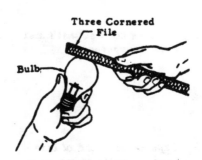

Three Cornered File

Bulb

1. File a small hole in the top of the bulb.

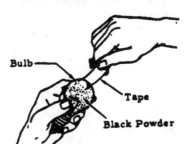

Bulb

2. Fill the bulb with Black Powder and tape the hole closed.

Tape

Black Powder

195

FUSE IGNITER FROM BOOK MATCHES

A simple, reliable fuse igniter can be made from paper book matches.

Material Required

Paper book matches.
Adhesive or friction tape.
Fuse cord (improvised or
   commercial).
Pin or small nail.

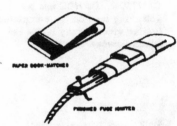

Procedure

1. Remove the staple(s) from match book and separate matches from cover.

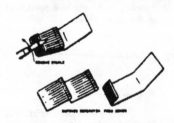

2. Cut fuse cord so that inner core is exposed.

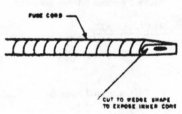

3. Tape exposed end of fuse cord in center of one row of matches.

4. Fold matches over fuse and tape.

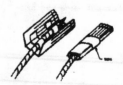

5. Shape the cover into a tube with the striking surface on the inside and tape. Make sure the edges of the cover at the striking end are butted. Leave cover open at opposite end for insertion of the matches.

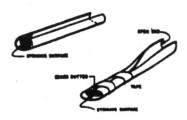

6. Push the taped matches with fuse cord into the tube until the bottom ends of the matches are exposed about 3/4 inch (2 cm).

7. Flatten and fold the open end of the tube so that it laps over about 1 inch (2-1/2 cm); tape in place.

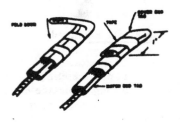

8. Push pin or small nail through matches and fuse cord. Bend end of pin or nail.

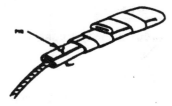

Method of Use

To light the fuse cord, the igniter is held by both hands and pulled sharply or quickly.

CAUTION

Store matches and completed fuse igniters in moistureproof containers such as plastic or rubber type bags until ready for use. Damp or wet paper book matches will not ignite. Fuse lengths should not exceed 12 in. (30 cm) for easy storage. These can be spliced to main fuses when needed.

197

DELAY IGNITER FROM CIGARETTE

A simple and economical time delay can be made with a common cigarette.

Materials Required

Cigarette.
Paper match.
String (shoelace or similar cord).
Fuse cord (improvised or commercial).

Procedure

←CUT SO INNER CORE IS EXPOSED

1. Cut end of fuse cord to expose inner core.

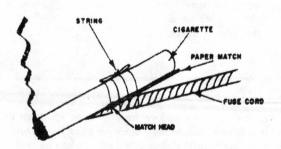

STRING   CIGARETTE   PAPER MATCH   FUSE CORD   MATCH HEAD

198   2. Light cigarette in normal fashion. Place a paper match so that the head is over exposed end of fuse cord and tie both to the side of the burning cigarette with string.

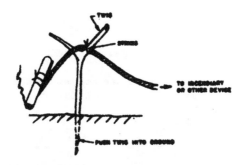

3. Position the burning cigarette with fuse so that it burns freely. A suggested method is to hang the delay on a twig.

NOTE

Common dry cigarettes burn about 1 inch every 7 or 8 minutes in still air. If the fuse cord is placed 1 inch from the burning end of a cigarette a time delay of 7 or 8 minutes will result.

Delay time will vary depending upon type of cigarette, wind, moisture, and other atmospheric conditions.

To obtain accurate delay time, a test run should be made under "use" conditions.

WATCH DELAY TIMER

A time delay device for use with electrical firing circuits can be made by using a watch with a plastic crystal.

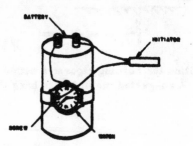

Material and Equipment Required

Watch with plastic crystal.
Small clean metal screw.
Battery.
Connecting wires.
Drill or nail.

Procedure

1. If watch has a sweep or large second hand, remove it. If delay time of more than one hour is required, also remove the minute hand. If hands are painted, carefully scrape paint from contact edge with knife.

2. Drill a hole through the crystal of the watch or pierce the crystal with a heated nail. The hole must be small enough that the screw can be tightly threaded nto it.

3. Place the screw in the hole and turn down as far as possible without making contact with the face of the watch. If screw has a pointed tip, it may be necessary to grind the tip flat.

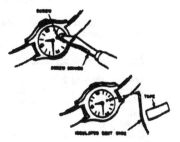

If no screw is available, pass a bent stiff wire through the hole and tape to the crystal.

IMPORTANT: Check to make sure hand of watch cannot pass screw or wire without contacting it.

How to Use

1. Set the watch so that a hand will reach the screw or wire at the time you want the firing circuit completed.

2. Wind the watch.

3. Attach a wire from the case of the watch to one terminal of the battery.

4. Attach one wire from an electric initiator (blasting cap, squib, or alarm device) to the screw or wire on the face of the watch.

5. After thorough inspection is made to assure that the screw or the wire connected to it is not touching the face or case of the watch, attach the other wire from the initiator to the second terminal of the battery.

CAUTION

Follow step 5 carefully to prevent premature initiation.

NO-FLASH FUSE IGNITER

A simple no-flash fuse igniter can be made from common pipe fittings.

MATERIAL REQUIRED:

1/4 in. (6mm) Pipe Cap
Solid 1/4 in. (6mm) Pipe Plug
Flat head nail about 1/16 in.
  (1 1/2 mm) in diameter
Hand Drill
Common "Strike Anywhere"
  Matches
Adhesive Tape

PROCEDURE:

1. Screw the pipe plug tightly into the pipe cap.

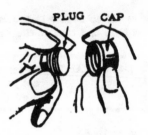

PLUG   CAP

2. Drill hole completely through the center of the plug and cap large enough that the nail fits loosely.

PLUG

DRILL   CAP

3. Enlarge the hole in the plug except for the last 1/8 in. (3 mm) so that the fuse cord will just fit.

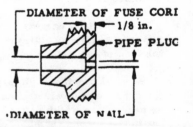

DIAMETER OF FUSE CORI
1/8 in.
PIPE PLUG

DIAMETER OF NAIL

4. Remove the plug from the cap and push the flat head nail through the hole in the cap from the inside.

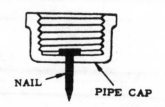

NAIL

PIPE CAP

5. Cut the striking tips from approximately 10 strike-anywhere matches. Place match tips inside pipe cap and screw plug in finger tight.

<u>HOW TO USE:</u>

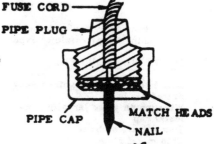

FUSE CORD

PIPE PLUG

PIPE CAP — MATCH HEADS — NAIL

1. Slide the fuse cord into the hole in the pipe plug.

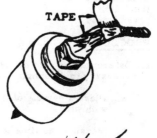

TAPE

2. Tape igniter to fuse cord.

3. Tap point of nail on a hard surface to ignite the fuse.

DRIED SEED TIMER

A time delay device for electrical firing circuits can be made using the principle of expansion of dried seeds.

MATERIEL REQUIRED:

Dried peas, beans or other dehy-
drated seeds
Wide mouth glass jar with non-
metal cap
Two screws or bolts
Thin metal plate
Hand drill
Screwdriver

PROCEDURE:

1. Determine the rate of rise of the dried seeds selected. This is necessary to determine delay time of the timer.

    a. Place a sample of the dried seeds in the jar and cover with water.

    b. Measure the time it takes for the seeds to rise a given height. Most dried seeds increase 50% in one to two hours.

2. Cut a disc from thin metal plate. Disc should fit loosely inside the jar.

NOTE: If metal is painted, rusty or otherwise coated, it must be scraped or sanded to obtain a clean metal surface.

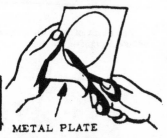

METAL PLATE

3. Drill two holes in the cap of the jar about 2 inches apart. Diameter of holes should be such that screws or bolts will thread tightly into them. If the jar has a metal cap or no cap, a piece of wood or plastic (NOT METAL) can be used as a cover.

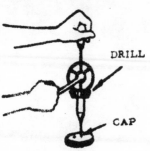

DRILL

CAP

4. Turn the two screws or bolts through the holes in the cap. Bolts should extend about one in. (2 1/2 cm) into the jar.

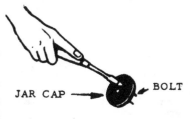

IMPORTANT: Both bolts must extend the same distance below the container cover.

JAR CAP → ← BOLT

5. Pour dried seeds into the container. The level will depend upon the previously measured rise time and the desired delay.

6. Place the metal disc in the jar on top of the seeds.

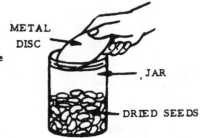

METAL DISC

JAR

DRIED SEEDS

HOW TO USE:

1. Add just enough water to completely cover the seeds and place the cap on the jar.

2. Attach connecting wires from the firing circuit to the two screws on the cap.

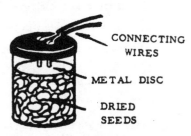

CONNECTING WIRES

METAL DISC

DRIED SEEDS

Expansion of the seeds will raise the metal disc until it contacts the screws and closes the circuit.

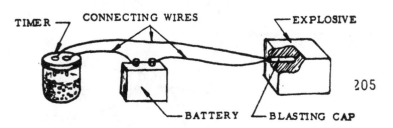

TIMER — CONNECTING WIRES — EXPLOSIVE

205

BATTERY — BLASTING CAP

FUSE CORDS

These fuse cords are used for igniting propellants and incendiaries or, with a non-electric blasting cap, to detonate explosives.

FAST BURNING FUSE

The burning rate of this fuse is approximately 40 in. (100 cm) per minute.

MATERIAL REQUIRED:

Soft Cotton String	Potassium Nitrate (Saltpeter) 25 parts
Fine Black Powder ---- or	Charcoal 3 parts
Piece of round stick	Sulphur 2 parts
Two pans or dishes	

PROCEDURE:

1. Moisten fine Black Powder to form a paste or prepare a substitute as follows:

    a. Dissolve Potassium Nitrate in an equal amount of water.

    b. Pulverize charcoal by spreading thinly on a hard surface and rolling the round stick over it to crush to a fine powder.

    c. Pulverize sulphur in the same manner.

    d. Dry mix sulphur and charcoal.

    e. Add Potassium Nitrate solution to the dry mix to obtain a thoroughly wet paste.

2. Twist or braid three strands of cotton string together.

STRING   NAIL

BOARD

3. Rub paste mixture into twisted string with fingers and allow to dry. BLACK POWDER PASTE

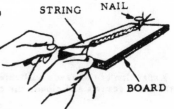

4. Check actual burning rate of fuse by measuring the time it takes for a known length to burn. This is used to determine the length needed for a desired delay time. If 5 in. (12 1/2 cm) burns for 6 seconds, 50 in. (125 cm) of fuse cord will be needed to obtain a one minute (60 second) delay time.

SLOW BURNING FUSE

The burning rate of this fuse is approximately 2 in. (5 cm) per minute.

MATERIAL REQUIRED:

Cotton String or 3 Shoelaces
Potassium Nitrate or Potassium Chlorate
Granulated Sugar

PROCEDURE:

1. Wash cotton string or shoelaces in hot soapy water; rinse in fresh water.

2. Dissolve 1 part Potassium Nitrate or Potassium Chlorate and 1 part granulated sugar in 2 parts hot water.

3. Soak string or shoelaces in solution.

4. Twist or braid three strands of string together and allow to dry.

5. Check actual burning rate of the fuse by measuring the time it takes for a known length to burn. This is used to determine the length needed for the desired delay time. If 2 in. (5 cm) burns for 1 minute, 10 in. (25 cm) will be needed to obtain a 5 minute delay.

NOTE: The last few inches of this cord (the end inserted in the material to be ignited) should be coated with the fast burning Black Powder paste if possible. This must be done when the fuse is used to ignite a blasting cap.

REMEMBER: The burning rate of either of these fuses can vary greatly. Do Not Use for ignition until you have checked their burning rate.

CLOTHESPIN TIME DELAY SWITCH

A 3 to 5 minute time delay switch can be made from the clothespin switch (Section VII, No. 1) and a cigarette. The system can be used for initiation of explosive charges, mines, and booby traps.

MATERIAL REQUIRED:

Spring type clothespin
Solid or stranded copper wire about 1/16 in. (2 mm) in diameter (field or
 bell wire is suitable)
Fine string, about 6 inches in length
Cigarette
Knife

PROCEDURE:

1. Strip about 4 inches (10 cm) of insulation from the ends of 2 copper wires. Scrape copper wires with pocket knife until metal is shiny.

2. Wind one scraped wire tightly on one jaw of the clothespin, and the other wire on the other jaw so that the wires will be in contact with each other when the jaws are closed.

3. Measuring from tip of cigarette, measure a length of cigarette that will correspond to the desired delay time. Make a hole in cigarette at this point, using wire or pin.

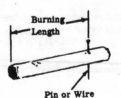

Burning Length

Pin or Wire

NOTE: Delay time may be adjusted by varying the burning length of the cigarette. Burning rate in still air is approximately 7 minutes per inch (2.5 cm). Since this rate varies with environment and brand of cigarette, it should be tested in each case if accurate delay time is desired.

4. Thread string through hole in cigarette.

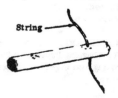

String

5. Tie string around rear of clothespin, 1/8 inch or less from end. The clothespin may be notched to hold the string in place.

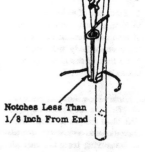

Notches Less Than 1/8 Inch From End

NOTE: The string must keep the rear end of the clothespin closed so that the jaws stay open and no contact is made between the wires.

HOW TO USE:

Suspend the entire system vertically with the cigarette tip down. Light tip of cigarette. Switch will close and initiation will occur when the cigarette burns up to and through the string.

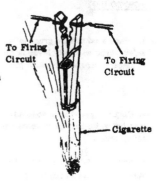

To Firing Circuit

To Firing Circuit

Cigarette

NOTE: Wires to the firing circuit must not be pulled taut when the switch is mounted. This could prevent the jaws from closing.

TIME DELAY GRENADE

This delay mechanism makes it possible to use an ordinary grenade as a time bomb.

MATERIAL REQUIRED:

Grenade
Fuse Cord

IMPORTANT: Fuse cord must be the type that burns completely. Slow burning improvised fuse cord (Section VI, No. 7) is suitable. Safety fuse is not satisfactory, since its outer covering does not burn.

PROCEDURE:

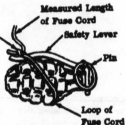

1. Bend end of safety lever upward to form a hook. Make a single loop of fuse cord around the center of the grenade body and safety lever. Tie a knot of the non-slip variety at the safety lever.

NOTE: The loop must be tight enough to hold the safety lever in position when the pin is removed.

2. Measuring from the knot along the free length of the fuse cord, measure off a length of fuse cord that will give the desired delay time. Cut off the excess fuse cord.

HOW TO USE:

1. Place hand around grenade and safety lever so safety lever is held in place. Carefully remove pin.

2. Emplace grenade in desired location while holding grenade and safety lever.

3. Very carefully remove hand from grenade and safety lever, making sure that the fuse cord holds the safety lever in place.

CAUTION: If loop and knot of fuse cord do not hold for any reason and the safety lever is released, the grenade will explode after the regular delay time.

4. Light free end of fuse cord.

CAN-LIQUID TIME DELAY

A time delay device for electrical firing circuits can be made using a can and liquid.

MATERIAL REQUIRED:

Can
Liquid (water, gasoline, etc.)
Small block of wood or any material that will float on the liquid used
Knife
2 pieces of solid wire, each piece 1 foot (30 cm) or longer

PROCEDURE:

1. Make 2 small holes at opposite sides of the can very close to the top.

2. Remove insulation from a long piece of wire for a distance a little greater than the diameter of the can.

3. Secure the wire in place across the top of the can by threading it through the holes and twisting in place, leaving some slack. Make loop in center or wire. Be sure a long piece of wire extends from one end of the can.

FOR OFFICIAL USE ONLY

4. Wrap a piece of insulated wire around the block of wood. Scrape insulation from a small section of this wire and bend as shown so that wire contacts loop before wood touches bottom of container. Thread this wire through the loop of bare wire.

5. Make a very small hole (pinhole) in the side of the container. Fill container with a quantity of liquid corresponding to the desired delay time. Since the rate at which liquid leaves the can depends upon weather conditions, liquid used, size of hole, amount of liquid in the container, etc., determine the delay time for each individual case. Delays from a few minutes to many hours are possible. Vary time by adjusting liquid level, type of liquid (water, oil) and hole size.

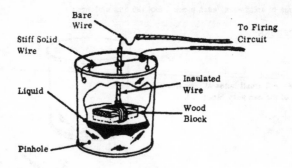

HOW TO USE:

1. Fill can with liquid to the same level as during experimental run (step 5 above). Be sure that wooden block floats on liquid and that wire is free to move down as liquid leaves container.

2. Connect wires to firing circuit.

NOTE: A long term delay can be obtained by placing a volatile liquid (gasoline, ether, etc.) in the can instead of water and relying on evapo-ration to lower the level. Be sure that the wood will float on the liquid used. DO NOT MAKE PINHOLE IN SIDE OF CAN!

FOR OFFICIAL USE ONLY

SHORT TERM TIME DELAY FOR GRENADE

A simple modification can produce delays of approximately 12 seconds for grenades when fired from Grenade Launchers (Section IV, No. 5).

MATERIAL REQUIRED:

Grenade
Nail
Knife }
Pliers } may not be needed
Safety fuse

NOTE: Any safety or improvised fuse may be used. However, since different time delays will result, determine the burning rate of the fuse first.

PROCEDURE:

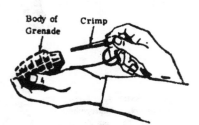

Body of   Crimp
Grenade

1. Unscrew fuse mechanism from body of grenade and remove. Pliers may have to be used.

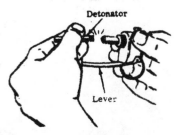

Detonator

2. Carefully cut with knife or break off detonator at crimp and save for later use.

Lever

CAUTION: If detonator is cut or broken below the crimp, detonation may occur and severe injuries could result.

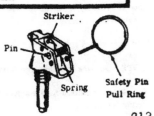

Striker

3. Remove safety pin pull ring and lever, letting striker hit the primer. Place fuse mechanism aside until delay fuse powder mix in mechanism is completely burned.

Pin

Spring

Safety Pin
Pull Ring

213

4. Remove pin, spring, and striker.

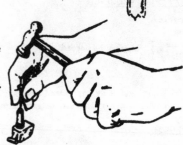

Primer

Fuse Mechanism
(Pin, Spring and
Striker Removed)

5. Remove primer from fuse
mechanism by pushing nail
through <u>bottom</u> end of primer
hole and tapping with hammer.

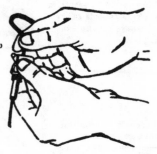

6. Insert safety fuse through top
of primer hole. Enlarge hole if
necessary. The fuse should go
completely through the hole.

7. Insert fuse into detonator and
tape it securely to modified fuse
mechanism.

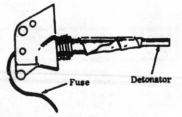

Fuse   Detonator

NOTE: Be sure that fuse rests firmly against detonator at all times.

8. Screw modified fuse mechanism back into grenade. Grenade is now
ready for use.

214

NOTE: If time delay is used for
Improvised Grenade Launchers
(Section IV, No. 5) -

1. Wrap tape around safety
   fuse.

2. Securely tape fuse to
   grenade.

3. Load grenade in launcher.
   Grenade will explode in
   approximately 12 seconds
   after safety fuse burns up
   to bottom of grenade.

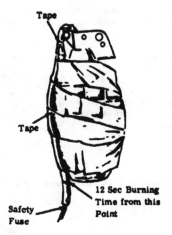

Tape

Tape

12 Sec Burning
Time from this
Point

Safety
Fuse

LONG TERM TIME DELAY FOR GRENADE

A simple modification can produce delays of approximately 20 seconds for grenades when fired from Grenade Launchers (Section IV, No. 5).

MATERIAL REQUIRED:

Grenade
Nail
"Strike-anywhere" matches, 6 to 8
Pliers (may not be needed)
Knife or sharp cutting edge
Piece of wood
Safety fuse

NOTE: Any safety or improvised fuse may be used. However, since different time delays will result, determine the burning rate of the fuse first.

PROCEDURE:

1. Unscrew fuse mechanism from body of grenade and remove. Pliers may have to be used.

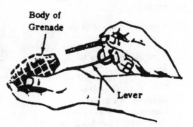

2. Insert nail completely through safety hole (hole over primer).

3. Carefully remove safety pin pull ring and lever, and allow striker to hit nail.

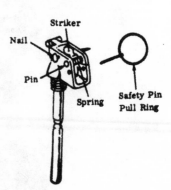

FOR OFFICIAL USE ONLY

CAUTION: If for any reason, striker should hit primer instead of nail, detonator will explode after (4-5 sec.) delay time.

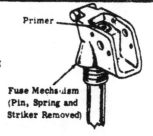

Primer

4. Push pin out and remove spring and striker. Remove nail.

Fuse Mechanism (Pin, Spring and Striker Removed)

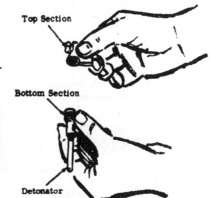

Top Section

5. Carefully remove top section of fuse mechanism from bottom section by unscrewing. Pliers may have to be used.

CAUTION: Use extreme care - sudden shock may set off detonator.

Bottom Section

Detonator

6. Fire primer by hitting nail placed against top of it. Remove fired primer (same as procedure 5 of Section VI, No. 11).

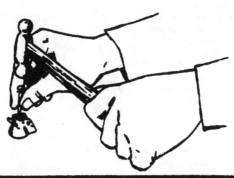

CAUTION: Do not hold assembly in your hand during above operation, as serious burns may result.

217

7. Scrape delay fuse powder with a sharpened stick. Loosen about 1/4 in. (6 mm) of powder in cavity.

8. Cut off tips (not whole head) of 6 "strike-anywhere" matches with sharp cutting edge. Drop them into delay fuse hole.

"Strike-Anywhere"   Head
Match Tip

9. Place safety fuse in delay fuse hole so that it is flush against the match tips.

IMPORTANT: Be sure fuse remains flush against the match tips at all times.

10. Thread fuse through primer hole. Enlarge hole if necessary. Screw modified fuse mechanism back together. Screw combination back into grenade. Grenade modification is now ready for use. Light fuse when ready to use.

NOTE: If time delay is used for
Improvised Grenade Launchers
(Section IV, No. 5) -

1. Wrap tape around safety
   fuse.

2. Securely tape fuse to
   grenade

3. Load grenade in launcher.
   Grenade will explode in
   approximately 20 seconds
   after safety fuse burns up
   to bottom of grenade.

Tape

Tape

Safety
Fuse

20 Sec Burning
Time from this
Point

FOR OFFICIAL USE ONLY

DETONATOR

Detonators (blasting caps) can be made from a used small arms cartridge case and field manufactured explosives. Detonators are used to initiate secondary high explosives (C-4, TNT, etc.).

MATERIAL REQUIRED:	SOURCE:
Primary explosive	See table
Booster explosive	RDX (Section I, No. 15) or Picric Acid (Section I, No. 21)
Improvised scale	Section VII, No. 8
Used cartridge case	.22 caliber or larger
Fuse, 12 in. long	
Round wooden stick (small enough just to fit in the neck of the cartridge case)	
Drill or knife	
Long nail with sharpened end	
Vise	
Improvised loading fixture	

PROCEDURE:

1. Remove fired primer from a used cartridge case using a sharpened nail. (See Section III, No. 5.)

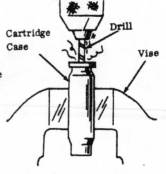

2. If necessary, open out flash hole in the primer pocket using a drill or knife. Make it large enough to receive fuse.

3. Place one end of fuse in the flash hole and extend it through the case until it becomes exposed at the open end. Knot this end and then pull fuse in cartridge case thus preventing fuse from falling out.

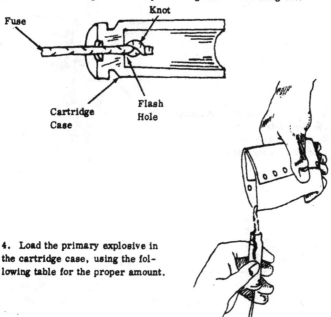

Knot

Fuse

Cartridge
Case

Flash
Hole

4. Load the primary explosive in the cartridge case, using the following table for the proper amount.

Primary Explosive	Primary Explosive Source	Minimum Weight*
Lead Picrate**	Section I, No. 20	3 grams (3 Handbook Pages)
TACC (Tetramminecopper Chlorate)	Section I, No. 16	1 gram (1 Handbook Page)
DDNP (Diazodinitrophenol)	Section I, No. 19	0.5 gram (1/2 Handbook Page)
Mercury Fulminate	Section I, No. 24	0.75 gram (3/4 Handbook Page)
HMTD	Section I, No. 17	
Double Salts	Section I, No. 22	

* See Section VII, No. 8 for details on improvised scale.
** .22 Cal. cartridge case cannot be used with lead picrate as there is not enough volume to contain the explosive train.

221

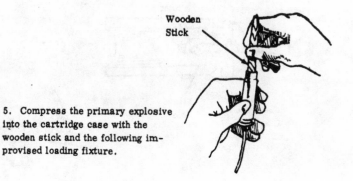

Wooden
Stick

5. Compress the primary explosive
into the cartridge case with the
wooden stick and the following improvised loading fixture.

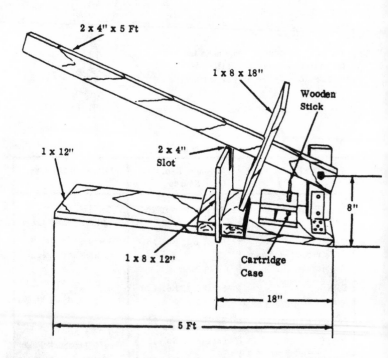

2 x 4" x 5 Ft

1 x 8 x 18"

Wooden
Stick

1 x 12"

2 x 4"
Slot

8"

1 x 8 x 12"

Cartridge
Case

18"

5 Ft

CAUTION: The primary explosive is shock and flame sensitive.

NOTE: Tamping is not needed when TACC is used.

222

6. Add one gram of booster explosive. The booster can be RDX (Section I, No. 15), or Picric Acid (Section I, No. 21).

7. Compress the booster explosive into the cartridge case with wooden stick and the loading fixture.

8. If the case is not full, fill the remainder with the secondary explosive to be detonated.

> CAUTION: Detonator has considerably more power than a military blasting cap and should be handled carefully.

CLOTHESPIN SWITCH

A spring type clothespin is used to make a circuit closing switch to actuate explosive charges, mines, booby traps
and alarm systems.

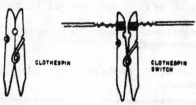

Material Required

Spring type clothespin.
Solid copper wire -- 1/16 in. (2 mm) in diameter.
Strong string on wire.
Flat piece of wood (roughly 1/8 x 1" x 2").
Knife.

Procedure

1. Strip four in. (10 cm) of in-
    sulation from the ends of 2
    solid copper wires. Scrape
    copper wires with pocket
    knife until metal is shiny.

2. Wind one scraped wire
    tightly on one jaw of the
    clothespin, and the other
    wire on the other jaw.

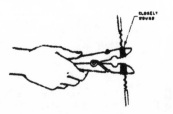

3. Make a hole in one end of
    the flat piece of wood using
    a knife, heated nail or drill
    Tie strong string or wire
    through the hole.

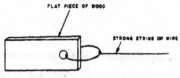

5. Place flat piece of wood between jaws of the clothespin
   switch.

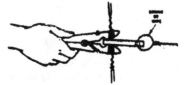

Basic Firing Circuit

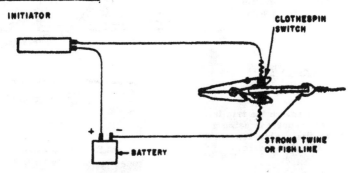

When the flat piece of wood is removed by pulling the
string, the jaws of the clothespin will close completing the
circuit.

CAUTION

Do not attach the battery until the switch and
trip wire have been emplaced and examined. Be
sure the flat piece of wood is separating the jaws
of the switch.

A Method of Use

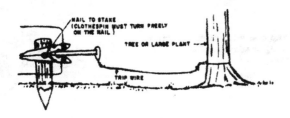

MOUSETRAP SWITCH

A common mousetrap can be used to make a circuit closing switch for electrically initiated explosives, mines and boobᵞ traps.

<u>MATERIEL REQUIRED:</u>

Mousetrap
Hacksaw or File
Connecting wires

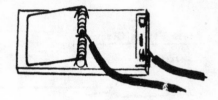

<u>PROCEDURE:</u>

1. Remove the trip lever from the mousetrap using a hacksaw or file. Also remove the staple and holding wire.

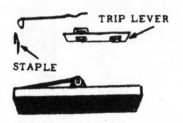

TRIP LEVER

STAPLE

2. Retract the striker of the mousetrap and attach the trip lever across the end of the wood base using the staple with which the holding wire was attached.

NOTE: If the trip lever is not made of metal, a piece of metal of approximately the same size should be used.

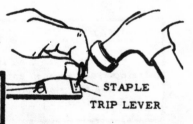

STAPLE
TRIP LEVER

3. Strip one in. (2 1/2 cm) of insulation from the ends of 2 connecting wires.

4. Wrap one wire tightly around the spring loaded striker of the mousetrap.

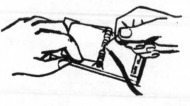

5. Wrap the second wire around some part of the trip lever or piece of metal.

CONNECTING WIRES

NOTE: If a soldering iron is available solder both of the above wires in place.

HOW TO USE:

This switch can be used in a number of ways -- one typical method is presented here.

The switch is placed inside a box which also contains the explosive and batteries. The spring loaded striker is held back by the lid of the box and when the box is opened the circuit is closed.

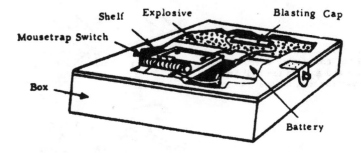

Shelf   Explosive   Blasting Cap

Mousetrap Switch

Box

Battery

FLEXIBLE PLATE SWITCH

This pressure sensitive switch is used for initiating emplaced mines and explosives.

MATERIAL REQUIRED:

Two flexible metal sheets
  one approximately 10 in. (25 cm) square
  one approximately 10 in. x 8 in.(20 cm)
Piece of wood 10 in. square by 1 in. thick
Four soft wood blocks 1 in.x 1 in.x 1/4 in.
Eight flat head nails, 1 in. long
Connecting wires
Adhesive tape

PROCEDURE:

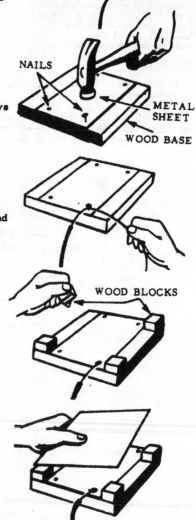

1. Nail 10 in. x 8 in. metal sheet to 10 in. square piece of wood so that 1 in. of wood shows on each side of metal. Leave one of the nails sticking up about 1/4 in.

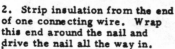

NAILS

METAL SHEET

WOOD BASE

2. Strip insulation from the end of one connecting wire. Wrap this end around the nail and drive the nail all the way in.

WOOD BLOCKS

3. Place the four wood blocks on the corners of the wood base.

4. Place the 10 in. square flexible metal sheet so that it rests on the blocks in line with the wood base.

228

5. Drive four nails through the metal sheet and the blocks to fasten to the wood base. A second connecting wire is attached to one of the nails as in Step 2.

6. Wrap adhesive tape around the edges of the plate and wood base. This will assure that no dirt or other foreign matter will get between the plates and prevent the switch from operating.

TAPE

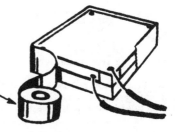

HOW TO USE:

The switch is placed in a hole in the path of expected traffic and covered with a thin layer of dirt or other camouflaging material. The mine or other explosive device connected to the switch can be buried with the switch or emplaced elsewhere as desired.

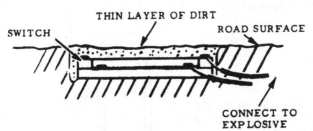

THIN LAYER OF DIRT

SWITCH   ROAD SURFACE

CONNECT TO EXPLOSIVE

When a vehicle passes over the switch, the two metal plates make contact closing the firing circuit.

229

METAL BALL SWITCH

This switch will close an electric circuit when it is tipped in any direction. It can be used alone for booby traps or in combination with another switch or timer as an anti-disturbance switch.

MATERIAL REQUIRED:

Metal Ball 1/2" (1 1/4 cm)
  diameter (see Note)
Solid copper wire 1/16" (1/4 cm)
  diameter
Wood block 1" (2 1/2 cm) square
  by 1/4" thick
Hand drill
Connecting wires
Soldering iron & solder

NOTE: If other than a 1/2" diameter ball is used, other dimensions must be changed so that the ball will rest in the center hole of the block without touching either of the wires.

PROCEDURE:

1. Drill four 1/16" holes and one 1/8" hole through the wood block as shown.

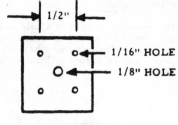

1/2"

1/16" HOLE
1/8" HOLE

2. Form two "U" shaped pieces from 1/16" copper wire to the dimensions shown.

ONE 1" HIGH
ONE 1-1/2" HIGH

3/4"

3. Wrap a connecting wire around one leg of each "U" at least 1/4" from the end and solder in place.

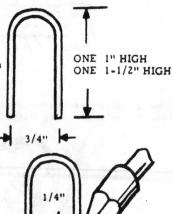

1/4"

230

4. Place metal ball on block so that it rests in the center hole.

5. Insert the ends of the small "U" into two holes in the block. Insert large "U" into the remaining two holes.

CAUTION: Make sure that the metal ball does not touch either "U" shaped wire when the switch is standing on its base. If the ball does touch, bend wires outward slightly.

HOW TO USE:

Mount switch vertically and connect in electrical firing circuit as with any other switch. When tipped in any direction it will close the circuit.

CAUTION: Switch must be mounted vertically and not disturbed while completing connections.

ALTIMETER SWITCH

This switch is designed for use with explosives placed on aircraft. It will close an electrical firing circuit when an altitude of approximately 5000 ft (1-1/2 KM) is reached.

MATERIAL REQUIRED:

Jar or tin can
Thin sheet of flexible plastic or waxed paper
Thin metal sheet (cut from tin can)
Adhesive Tape
Connecting Wires

PROCEDURE:

1. Place sheet of plastic or waxed paper over the top of the can or jar and tape tightly to sides of container.

NOTE: Plastic sheet should not be stretched tight. A small depression should be left in the top.

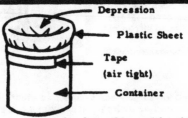

Depression

Plastic Sheet

Tape
(air tight)

Container

2. Cut two contact strips from thin metal and bend to the shapes shown.

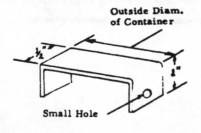
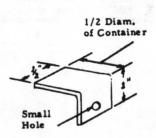

Outside Diam.
of Container

Small Hole

1/2 Diam.
of Container

Small
Hole

3. Strip insulation from the ends of two connecting wires. Attach one wire to each contact strip.

NOTE: If a soldering iron is available solder wires in place.

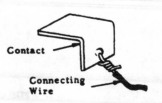

Contact

Connecting
Wire

4. Place contact strips over container so that the larger contact is above the smaller with a very small clearance between the two.

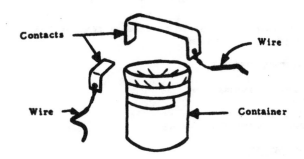

Contacts

Wire

Wire

Container

5. Securely tape contact strips to sides of container.

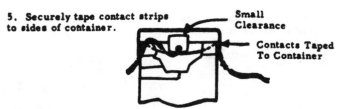

Small Clearance

Contacts Taped To Container

HOW TO USE:

1. Connect the altimeter switch in an explosive circuit the same as any switch.

2. Place the explosive package on airplane. As the plane rises the air inside the container will expand. This forces the plastic sheet against the contacts closing the firing circuit.

NOTE: The switch will not function in a pressurized cabin. It must be placed in some part of the plane which will not be pressurized.

FOR OFFICIAL USE ONLY

PULL-LOOP SWITCH

This switch will initiate explosive charges, mines, and booby traps when the trip wire is pulled.

MATERIAL REQUIRED:

2 lengths of insulated wire
Knife
Strong string or cord
Fine thread that will break easily

PROCEDURE:

1. Remove about 2 inches of insulation from one end of each length of wire. Scrape bare wire with knife until metal is shiny.

2. Make a loop out of each piece of bare wire.

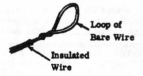

Loop of Bare Wire

Insulated Wire

3. Thread each wire through the loop of the other wire so the wires can slide along each other.

NOTE: The loops should contact each other when the two wires are pulled taut.

234

HOW TO USE:

1. Separate loops by about 2 inches. Tie piece of fine thread around wires near each loop Thread should be taut enough to support loops and wire, yet fine enough that it will break under a very slight pull.

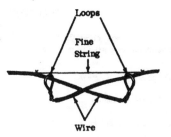

2. Fasten one wire to tree or stake and connect end to firing circuit.

3. Tie a piece of cord or string around the other piece of wire a few inches from the loop. Tie free end of cord around tree, bush, or stake. Connect the free end of the wire to the firing circuit. Initiation will occur when the tripcord is pulled.

CAUTION: Be sure that the loops do not contact each other when the wires are connected to the firing circuit.

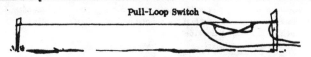

OTHER USES: The switch minus the fine thread may be used to activate a booby trap by such means as attaching it between the lid and a rigid portion of a box, between a door and a door jamb, and in similar manners.

KNIFE SWITCH

This device will close the firing circuit charges, mines, and booby traps when the trip wire is pulled or cut.

MATERIAL REQUIRED:

Knife or hack saw blade   Sturdy wooden board
6 nails   Wire
Strong string or light rope

PROCEDURE:

1. Place knife on board. Drive 2 nails into board on each side of knife handle so knife is held in place.

2. Drive one nail into board so that it touches blade of knife near the point.

3. Attach rope to knife. Place rope across path. Apply tension to rope, pulling knife blade away from nail slightly. Tie rope to tree, bush, or stake.

4. Drive another nail into board near the tip of the knife blade as shown below. Connect the two nails with a piece of conducting wire. Nail should be positioned so that it will contact the second nail when blade is pulled about 1 inch (2-1/2 cm) to the side.

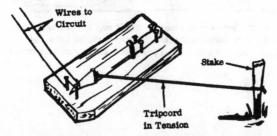

NOTE: Check position of nails to knife blade. The nails should be placed so that the knife blade will contact either one when the rope is pulled or released.

HOW TO USE:

Attach one wire from firing circuit to one of the nails and the other to the knife blade. The circuit will be completed when the tripcord is pulled or released.

IMPROVISED SCALE

This scale provides a means of weighing propellant and other items when conventional scales or balances are not available.

MATERIAL REQUIRED:

Pages from Improvised Munitions Handbook
Straight sticks about 1 foot (30 cm) long and 1/4 in. (5 mm) in diameter
Thread or fine string

PROCEDURE:

1. Make a notch about 1/2 in. (1 cm) from each end of stick. Be sure that the two notches are the same distance from the end of the stick.

2. Find the exact center of the stick by folding in half a piece of thread the same length as the stick and placing it alongside the stick as a ruler. Make a small notch at the center of the stick.

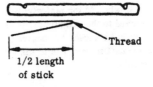

3. Tie a piece of thread around the notch. Suspend stick from branch, another stick wedged between rocks, or by any other means. Be sure stick is balanced and free to move.

NOTE: If stick is not balanced, shave or scrape a little off the heavy end until it does balance. Be sure the lengths of the arms are the same.

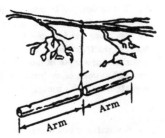

4. Make a container out of one piece of paper. This can be done by rolling the paper into a cylinder and folding up the bottom a few times.

5. Punch 2 holes at opposite sides of paper container. Suspend container from one side of stick.

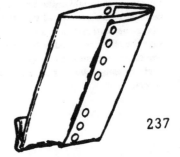

237

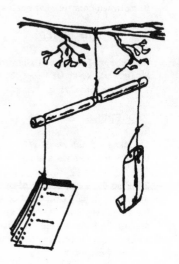

6. Count out the number of hand-book pages equal in weight to that of the quantity of material to be weighed. Each sheet of paper weighs about 1.3 grams (20 grains or .04 ounce). Suspend these sheets, <u>plus one</u>, to balance container on the other side of the scale.

7. Slowly add the material to be weighed to the container. When the stick is balanced, the desired amount of material is in the container.

8. If it is desired to weigh a quantity of material larger than that which would fit in the above container, make a container out of a larger paper or paper bag, and suspend from one side of the stick. Suspend handbook pages from the other side until the stick is balanced. Now place a number of sheets of handbook pages equal in weight to that of the desired amount of material to be weighed on one side, and fill the container with the material until the stick is balanced.

9. A similar method may be used to measure parts or percentage by weight. The weight units are unimportant. Suspend equal weight containers from each side of the stick. Bags, tin cans, etc. can be used. Place one material in one of the containers. Fill the other container with the other material until they balance. Empty and refill the number of times necessary to get the required parts by weight (e.g., 5 to 1 parts by weight would require 5 fillings of one can for one filling of the other).

ROPE GRENADE LAUNCHING TECIINIQUE

A method of increasing the distance a grenade may be thrown. Safety fuse is used to increase the delay time.

MATERIAL REQUIRED:

Hand grenade (Improvised pipe hand grenade, Section II, No. 1 may be used)
Safety fuse or fast burning Improvised Fuse, (Section VI, No. 7)
Light rope, cord, or string

PROCEDURE:

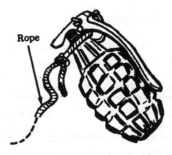

1. Tie a 4 to 6 foot (2 meter) length of cord to the grenade. Be sure that the rope will not prevent the grenade handle from coming off.

Rope

Note: If improvised grenade is used, tie cord around grenade near the end cap. Tape in place if necessary.

2. Tie a large knot in the other end of the cord for use as a handle.

3. Carefully remove safety pin from grenade, holding safety lever in place. Enlarge safety pin hole with point of knife, awl, or drill so that safety fuse will pass through hole.

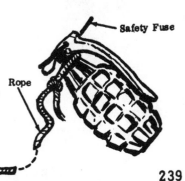

Safety Fuse

4. Insert safety fuse in hole. Be sure that safety fuse is long enough to provide a 10 second or more time delay. Slowly release safety lever to make sure fuse holds safety lever in place.

Rope

> **CAUTION:** If safety lever should be released for any reason, grenade will explode after regular delay time (4-5 sec.).

NOTE: If diameter of safety fuse is too large to fit in hole (Step 4), follow procedure and How to Use of Time Delay Grenade, Section VI, No. 9, instead of Steps 3 and 4 above.

HOW TO USE:

1. Light fuse.

2. Whirl grenade overhead, holding knot at end of rope, until grenade picks up speed (3 or 4 turns).

3. Release when sighted on target.

> **CAUTION:** Be sure to release grenade within 10 seconds after fuse is lit.

NOTE: It is helpful to practice first with a dummy grenade or a rock to improve accuracy. With practice, accurate launching up to 100 meters (300 feet) can be obtained.

BICYCLE GENERATOR POWER SOURCE

A 6 volt, 3 watt bicycle generator will set off one or two blasting caps (connected in series) or an igniter.

MATERIAL REQUIRED:

Bicycle generator (6 volts, 3 watt)
Copper wire
Knife

PROCEDURE:

1. Strip about 4 in. (10 cm) of coating from both ends of 2 copper wires. Scrape ends with knife until metal is shiny.

2. Connect the end of one wire to the generator terminal.

3. Attach the end of the second wire to generator case. This wire may be wrapped around a convenient projection, taped, or simply held against the case with the hand.

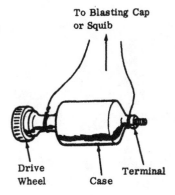

To Blasting Cap
or Squib

Drive
Wheel

Case

Terminal

HOW TO USE:

1. Connect free ends of wires to blasting cap or squib leads.

> CAUTION: If drive wheel is rotated, explosive may be set off.

2. Run the drive wheel firmly and rapidly across the palm of the hand to activate generator.

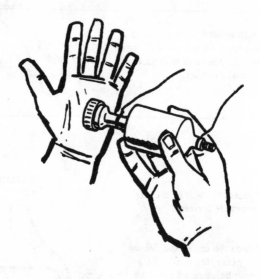

FOR OFFICIAL USE ONLY

AUTOMOBILE GENERATOR POWER SOURCE

An automobile generator can be used as a means of firing one blasting cap or igniter. (Improvised Igniter, Section V, No. 2, may be used.)

MATERIAL REQUIRED:

Automobile generator (6, 12, or 28 volts). (An alternator will not work.)
Copper Wire
Strong string or wire, about 5 ft. (150 cm) long and 1/16 in. (1-1/2 mm)
    in diameter
Knife
Small light bulb requiring same voltage as generator, (for example,
    bulb from same vehicle as generator).

PROCEDURE:

1. Strip about 1 in. (2-1/2 cm) of
coating from both ends of 3 copper
wires. Scrape ends with knife
until metal is shiny.

2. Connect the A and F terminals
with one piece of wire.

3. Connect a wire to the A
terminal. Connect another
to the G terminal.

NOTE: The F and G or C terminals may not be labeled; in this case,
connect wires as shown. The F terminal is usually smaller in size
than the C or G terminal.

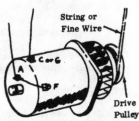

String or
Fine Wire

4. Wrap several turns of string or wire clockwise around the drive pulley.

Drive
Pulley

HOW TO USE:

1. Connect the free ends of the wires to the light bulb.

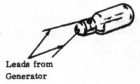

Leads from
Generator

Leads to Bulb
or Detonator

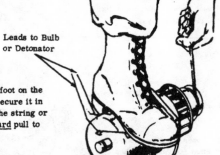

2. Place one foot on the generator to secure it in place. Give the string or wire a very hard pull to light the bulb.

NOTE: If not successful at first, rewind string and try again several times. After repeating this operation and the bulb still does not light, follow Step 4, "How to Use."

3. If light bulb lights, follow Steps 1 and 2 of above, "How to Use," connecting free ends of wires to blasting cap or igniter instead of to light bulb.

4. If light bulb does not light after several pulls, switch leads connected to F and G terminals. Repeat above "How to Use," Steps 1 to 3.

244

IMPROVISED BATTERY (SHORT LASTING)

This battery is powerful but must be used within 15 minutes after fabrication. One cell of this battery will detonate one blasting cap or one igniter. Two cells, connected in series, will detonate two of these devices and so on. Larger cells have a longer life as well as greater power.

MATERIALS	COMMON SOURCE
Water	
Sodium hydroxide (lye, solid or concentrated solution)	Soap manufacturing Disinfectants Sewer cleaner
Copper or brass plate about 4 in. (10 cm) square and 1/16 in. (2 mm) thick	
Aluminum plate or sheet, same size as copper plate	
Charcoal powder	
Container for mixing	
Knife	

One of the following:

Potassium permanganate, solid	Disinfectants Deodorants
Calcium hypochlorite, solid	Disinfectants Water treating chemicals Chlorine bleaches
Manganese dioxide (pyrolucite)	Dead dry-cell batteries

NOTE: Be sure sodium hydroxide solution is at least a 45% solution by weight. If not, boil off some of the water. If solid sodium hydroxide is available, dissolve some sodium hydroxide in about twice as much water (by volume).

<u>PROCEDURE</u>:

1. Scrape coating off both
ends of wires with knife
until metal is shiny.

2. Mix thoroughly (do not grind) approximately equal volumes of
powdered charcoal and <u>one</u> of the following: potassium permangenate,
calcium hypochlorite, or manganese dioxide. Add water until a very
thick paste is formed.

CAUTION: Avoid getting any of the ingredient on the skin or in the eyes.

3. Spread a layer of this
mixture about 1/8 in. (2 mm)
thick on the copper or brass
plate. Be sure mixture is
thick enough so that when
mixture is sandwiched be-
tween two metal plates,
the plates will not touch
each other at any point.

NOTE: If more power is required, prepare several plates as above.

<u>HOW TO USE</u>:

1. Just prior to use (no
more than 15 minutes),
carefully pour a small
quantity of sodium hydroxide solution over the mixture on each plate used.

CAUTION: If solution gets on skin, wash off immediately with water.

2. Place an aluminum plate on top of the mixture on each copper plate. Press firmly. Remove any excess that oozes out between the plates.

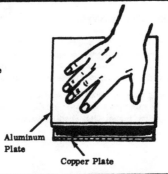

Aluminum Plate

Copper Plate

CAUTION: Be sure plates are not touching each other at any point.

3. If more than one cell is used, place the cells on top of each other so that unlike metal plates are touching.

Aluminum Plate

Copper Plate

Aluminum Plate

Copper Plate

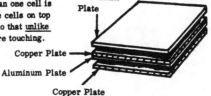

4. When ready to fire, clean plates with knife where connections are to be made. Connect one wire to the outer aluminum plate. This may be done by holding the wires against the plates or by hooking them through holes punched through plates. If wires are hooked through plates, be sure they do not touch mixture between plates.

Copper Wire

Aluminum Plate

Copper Plate

Copper Plate

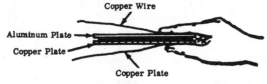

Copper Wire

Aluminum Plate

Copper Plate

Aluminum Plate

Copper Wire

Copper Plate

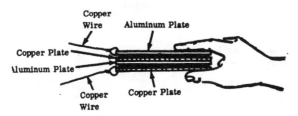

247

IMPROVISED BATTERY (2 HOUR DURATION)

This battery should be used within 2 hours and should be securely wrapped. Three cells will detonate one blasting cap or one igniter. Five cells, connected in series, will detonate two of these devices and so on. Larger cells have a longer life and will yield more power.

If depolarizing materials such as potassium permanganate or manganese dioxide cannot be obtained, ten cells without depolarizer, arranged as described below, (Step 4) will detonate one blasting cap.

MATERIALS	COMMON SOURCE
Water	
Ammonium chloride (sal ammoniac) (solid or concentrated solution)	Medicines Soldering fluxes Fertilizers Ice melting chemicals for roads
Charcoal powder	
Copper or brass plate about 4 in. (10 cm) square and 1/16 in. (2 mm) thick	
Aluminum plate same size as copper or brass plate	
Wax and paper (or waxed paper)	Candles
Wire, string or tape	
Container for mixing	
Knife	

One of the following:

Potassium permanganate, solid	Disinfectants Deodorants
Manganese dioxide	Dead dry batteries

NOTE: If ammonium chloride solution is not concentrated (at least 45% by weight) boil off some of the water.

PROCEDURE:

1. Mix thoroughly (do not grind) approximately equal volumes of powdered charcoal, ammonium chloride and one of the following: potassium
permanganate or manganese dioxide. Add water until a very thick paste
is formed. If ammonium chloride is in solution form, it may not be
necessary to add water.

2. Spread a layer of this mixture, about 1/8 in. (3 mm) thick,
on a clean copper or brass plate.
The layer must be thick enough
to prevent a second plate from
touching the copper plate when
it is pressed on top.

3. Press an aluminum plate very
firmly upon the mixture on the copper plate. Remove completely any
of the mixture that squeezes out
between the plates. The plates
must not touch.

Aluminum
Plate

Copper Plate

4. If more than one cell is desired:

a. Place one cell on top of
the other so that unlike
metal plates are touching.

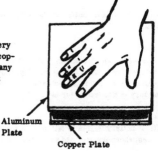

Aluminum
Plate

Copper Plate

Aluminum Plate

Copper Plate

249

b. Wrap the combined cells
in heavy waxed paper.
The waxed paper can be
made by rubbing candle
wax over one side of a
piece of paper. Secure
the paper around the
battery with string,
wire or tape. Expose
the top and bottom met-
al plates at one corner.

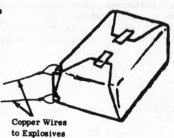

Copper Wires
to Explosives

<u>HOW TO USE</u>:

1. Scrape a few inches off each end of two wires with knife till metal is
shiny.

2. Clean plates with knife until metal is shiny where connections are
to be made.

3. Connect one wire from the explosive to a copper or brass plate and
the other wire to an aluminum plate. The connection can be made by
holding the wire against the plate. A permanent connection can be made
by hooking the wire through holes in the exposed corners of the plates.
The battery is now ready for use.

NOTE: If battery begins to fail after a few firings, scrape the plates
and wires where connections are made until metal is shiny.

ARMOR MATERIALS

The following table shows the amount of indigenous materials needed to stop ball type projectiles of the 5.56 mm, .30 caliber, and .50 caliber ammunition fired from their respective weapons at a distance of 10 feet (3 m).

INDIGENOUS MATERIAL	THICKNESS OF MATERIALS					
	Inches			Centimeters		
	5.56 mm	.30 cal 7.62 mm	.50 cal 12.70 mm	5.56 mm	.30 cal 7.62 mm	.50 cal 12.70 mm
Mild steel (structural)	$\frac{1}{2}$	$\frac{1}{2}$	$\frac{3}{4}$	$1\frac{1}{4}$	$1\frac{1}{4}$	2
Mild aluminum (structural)	1	1	2	$2\frac{1}{2}$	$2\frac{1}{2}$	5
Pine wood (soft)	14	22	32	36	56	82
Broken stones (cobble gravel)	3	4	11	8	11	28
Dry sand	4	5	14	11	13	36
Wet sand or earth	6	13	21	16	33	54

NOTE: After many projectiles are fired into the armor, the armor will break down. More material must be added.

Section 13. PRIMARY HIGH EXPLOSIVES

MERCURY FULMINATE   13–1

Description: Mercury fulminate is an initiating explosive, commonly appearing as white or gray crystals. It is extremely sensitive to initiation by heat, friction, spark or flame, and impact. It detonates when initiated by any of these means. It is pressed into containers, usually at 3000 psi, for use in detonators and blasting caps. However, when compressed at greater and greater pressure (up to 30,000 psi), it becomes "dead pressed." In this condition, it can only be exploded by another initial detonating agent. Mercury fulminate gradually becomes inert when stored continuously above 100° F. A dark-colored product of deterioration gives evidence of this effect. Mercury fulminate is stored underwater execpt when there is danger of freezing. Then it is stored under a mixture of water and alcohol.

Comments: This material was tested. It is effective.

References: TM 9–1900, Ammunition, General, page 59.
TM 9–1910, Military Explosives, page 98.

LEAD STYPHNATE   13–2

Description: Lead styphnate is an initiating explosive, commonly appearing in the form of orange or brown crystals. It is easily ignited by heat and static discharge but cannot be used to initiate secondary high explosives reliably. Lead styphnate is used as an igniting charge for lead azide and as an ingredient in priming mixtures for small arms ammunition. In these applications, it is usually mixed with other materials first and then pressed into a metallic container (detonators and primers). Lead styphnate is stored under water except when there is danger of freezing. Then it is stored under a mixture of water and alcohol.

Comments: This item was tested. It is effective.

References: TM 9–1900, Ammunition, General, page 59.
TM 9–1910, Military Explosives, page 107.

LEAD AZIDE   13–3

Description: Lead azide is an initiating explosive and is produced as a white to buff crystalline substance. It is a more efficient detonating agent than mercury fulminate and it does not decompose on long continued storage at moderately elevated temperatures. It is sensitive to both flame and impact but requires a layer of lead styphnate priming mixture to produce reliable initiation when it is used in detonators that are initiated by a firing pin or electrical energy. It is generally loaded into aluminum detonator housings and must not be loaded into housing of copper or brass because extremely sensitive copper azide can be formed in the presence of moisture.

Comments: This material was tested. It is effective.

References: TM 9–1900, Ammunition, General, page 60.
TM 9–1910, Military Explosives, page 103.

DDNP 13-4

Description: DDNP (diazodinitrophenol) is a primary high explosive. It is extensively used in commercial blasting caps that are initiated by black powder safety fuse. It is superior to mercury fulminate in stability but is not as stable as lead azide. DDNP is desensitized by immersion in water.

Comments: This material was tested. It is effective.

References: TM 9–1900, Ammunition, General, page 60.
TM 9–1910, Military Explosives, page 103.

Section 14. SECONDARY HIGH EXPLOSIVES

TNT 14-1

Description: TNT (Trinitrotoluene) is produced from toluene, sulfuric acid, and nitric acid. It is a powerful high explosive. It is well suited for steel cutting, concrete breaching, general demolition, and for underwater demolition. It is a stable explosive and is relatively insensitive to shock. It may be detonated with a blasting cap or by primacord. TNT is issued in 1-pound and ½-pound containers and 50-pounds to a wooden box.

Comments: This material was tested. It is effective. TNT is toxic and its dust should not be inhaled or allowed to contact the skin.

References: TM 9–1900, Ammunition, General, page 263.
FM 5–25, Explosives and Demolitions, page 3.

NITROSTARCH 14-2

Description: Nitrostarch is composed of starch nitrate, barium nitrate, and sodium nitrate. It is more sensitive to flame, friction, and impact than TNT but is less powerful. It is initiated by detonating cord. Nitrostarch is issued in 1-pound and 1½-pound blocks. The 1-pound packages can be broken into ¼-pound blocks. Fifty 1-pound packages and one hundred 1½-pound packages are packed in boxes.

Comments: This material was tested. It is effective.

Reference: TM 9–1900, Ammunition, General, page 263.

TETRYL 14-3

Description: Tetryl is a fine, yellow, crystalline material and exhibits a very high shattering power. It is commonly used as a booster in explosive trains. It is stable in storage. Tetryl is used in detonators. It is pressed into the bottom of the detonator housing and covered with a small priming charge of mercury fulminate or lead azide.

Comments: This material was tested. It is effective.

References: TM 9–1900, Ammunition, General, page 52.
TM 31–201–1, Unconventional Warfare Devices and Techniques, para 1509.

RDX 14–4

Description: RDX (cyclonite) is a white crystalline solid that exhibits very high shattering power. It is commonly used as a booster in explosive trains or as a main bursting charge. It is stable in storage, and when combined with proper additives, may be cast or press loaded. It may be initiated by lead azide or mercury fulminate.

Comments: This material was tested. It is effective.

References: TM 9–1900, Ammunition, General, page 52.
TM 31–201–1, Unconventional Warfare Devices and Techniques, para 1501.

NITROGLYCERIN 14–5

Description: Nitroglycerin is maufactured by treating glycerin with a nitrating mixture of nitric and sulfuric acid. It is a thick, clear to yellow-brownish liquid that is an extremely powerful and shock-sensitive high explosive. Nitroglycerin freezes at 56° F., in which state it is less sensitive to shock than in liquid form.

Comments: This material was tested. It is effective.

References: TM 9–1910, Military Explosives, page 123.
TM 31–201–1, Unconventional Warfare Devices and Techniques, para 1502.

COMMERCIAL DYNAMITE 14–6

Description: There are three principal types of commercial dynamite: straight dynamite, ammonia dynamite, and gelatin dynamite. Each type is further subdivided into a series of grades. All dynamites contain nitroglycerin in varying amounts and the strength or force of the explosive is related to the nitroglycerin content. Dynamites range in velocity of detonation from about 4000 to 23,000 feet per second and are sensitive to shock. The types and grades of dynamite are each used for specific purposes such as rock blasting or underground explosives. Dynamite is initiated by electric or nonelectric blasting caps. Although dynamites are furnished in a wide variety of packages, the most common unit is the ½ pound cartridge. Fifty pounds is the maximum weight per case.

Comments: This material was tested. It is effective.

References: TM 9-1900, Ammunition, General, page 265.

FM 5-25, Explosives and Demolitions, page 8.

MILITARY DYNAMITE 14-7

Description: Military (construction) dynamite, unlike commercial dynamite, does not absorb or retain moisture, contains no nitroglycerine, and is much safer to store, handle, and transport. It comes in standard sticks 1¼ inches in diameter by 8 inches long, weighing approximately ½ pound. It detonates at a velocity of about 20,000 feet per second and is very satisfactory for military construction, quarrying, and demolition work. It may be detonated with an electric or nonelectric military blasting cap or detonating cord.

Comments: This material was tested. It is effective.

References: FM 5-25, Explosives and Demolitions, page 7.

TM 9-1910, Military Explosives, page 204.

AMATOL 14-8

Description: Amatol is a high explosive, white to buff in color. It is a mixture of ammonium nitrate and TNT, with a relative effectiveness slightly higher than that of TNT alone. Common compositions vary from 80% ammonium nitrate and 20% TNT to 40 % ammonium nitrate and 60% TNT. Amatol is used as the main bursting charge in artillery shell and bombs. Amatol absorbs moisture and can form dangerous compounds with copper and brass. Therefore it should not be housed in containers of such metals.

Comments: This material was tested. It is effective.

References: FM 5-25, Explosives and Demolitions, page 7.

TM 9-1910, Military Explosives, page 182.

PETN 14-9

Description: PETN (pentaerythrite tetranitrate), the high explosive used in detonating cord, is one of the most powerful of military explosives, almost equal in force to nitroglycerine and RDX. When used in detonating cord, it has a detonation velocity of 21,000 feet per second and is relatively insensitive to friction and shock from handling and transportation.

Comments: This material was tested. It is effective.

References: FM 5-25, Explosives and Demolitions, page 7.

TM 9-1910, Military Explosives, page 135.

TM 31-201-1, Unconventional Warfare Devices and Techniques, para 1508.

BLASTING GELATIN   **14–10**

Description: Blasting gelatin is a translucent material of an elastic, jellylike texture and is manufactured in a number of different colors. It is considered to be the most powerful industrial explosive. Its characteristics are similar to those of gelatin dynamite except that blasting gelatin is more water resistant.

Comments: This material was tested. It is effective.

Reference: TM 9–1910, Military Explosives, page 204.

COMPOSITION B   **14–11**

Description: Composition B is a high-explosive mixture with a relative effectiveness higher than that of TNT. It is also more sensitive than TNT. It is composed of RDX (59%), TNT (40%), and wax (1%). Because of its shattering power and high rate of detonation, Composition B is used as the main charge in certain models of bangalore torpedoes and shaped charges.

Comments: This material was tested. It is effective.

References: FM 5–25, Explosives and Demolitions, page 7.
          TM 9–1900, Ammunition, General, page 57.
          TM 9–1910, Military Explosives, page 193.

COMPOSITION C4   **14–12**

Description: Composition C4 is a white plastic high explosive more powerful than TNT. It consists of 91% RDX and 9% plastic binder. It remains plastic over a wide range of temperatures (—70° F. to 170° F.), and is about as sensitive as TNT. It is eroded less than other plastic explosives when immersed under water for long periods. Because of its high detonation velocity and its plasticity, C4 is well suited for cutting steel and timber and for breaching concrete.

Comments: This material was tested. It is effective.

Reference: TM 9–1910, Military Explosives, page 204.

AMMONIUM NITRATE   **14–13**

Description: Ammonium nitrate is a white crystalline substance that is extremely water absorbent and is therefore usually packed in a sealed metal container. It has a low velocity of detonation (3600 fps) and is used primarily as an additive in other explosive compounds. When it is used alone, it must be initiated by a powerful booster or primer. It is only 55% as powerful as TNT, hence larger quantities are required to produce similar results.

Comments: This material was tested. It is effective.

   Caution: **Never use copper or brass containers because ammonium nitrate reacts with these metals.**

References: TM 9–1900, Ammunition, General, page 264.
          TM 9–1910, Military Explosives, page 119.

CPSIA information can be obtained
at www.ICGtesting.com
Printed in the USA
BVHW081600140920
588699BV00013B/810

9 781626 542686